Human Health Guide:

The Missing Truths They never Told You

PUBLISHED BY Ember Maple Editions

1

Table of contents

Introduction: The Great Health Deception

"The doctor of the future will give no medicine, but will interest his patients in the care of the human frame, in diet, and in the cause and prevention of disease."
– Thomas Edison

We live in a time where humanity has never had more access to health information, yet paradoxically, never before have so many people been chronically unwell. Heart disease, diabetes, autoimmune disorders, anxiety, obesity, cancer—these have become normalized, almost expected. Most people assume that deteriorating health is just part of aging, or bad luck, or a random shuffle in genetic cards. Few question the system itself. Fewer still look behind the glossy screens of pharmaceutical ads or beneath the surface of dietary guidelines. But you're here now, and that means you're ready to look deeper. This book is for you.

This is not a manual of conspiracy theories, nor a rejection of science or medicine. On the contrary—it is a deep reconnection to science in its purest, most human form: the pursuit of truth through observation, understanding, and natural balance. It is also an act of remembering. Remembering what has been lost in translation between nature and lab coat, between soil and supermarket shelf, between your body's ancient intelligence and the loud, often conflicting messages of the modern world.

The Age of Silent Sabotage

Let's begin with the uncomfortable reality: the systems we trust to protect our health are often designed to manage sickness, not to promote wellness. The modern healthcare industry is dominated by symptom management. This is not always due to malice, but it is often driven by incentives that have little to do with your vitality. Pharmaceutical companies are legally beholden to shareholders, not patients. Their profits rise not when you're well, but when you're dependent.

And yet we call it "healthcare."

What we really have is a disease-care system. A machine built to intervene when things break down, not to guide us toward wholeness. And while this system can be life-saving in emergencies, it falters miserably when it comes to chronic, lifestyle-driven conditions. Doctors are trained to diagnose and prescribe, but rarely to educate. Nutrition courses in medical schools are, at best, elective—and often funded by the very industries that profit from ultra-processed foods.

We've outsourced our health to a fragmented model that treats the body like a car with interchangeable parts, not a living, interconnected ecosystem. And this disconnection is not accidental. It is the result of decades—if not centuries—of compartmentalized thinking, political lobbying, and corporate influence.

You Were Never Meant to Be Sick

Let this sink in: the human body is not designed to fail in middle age. It is not meant to be inflamed, fatigued, depressed, and full of pain. These are not normal signs of aging. They are symptoms of a culture gone off course.

Your body is an intricate, adaptive, self-healing organism. It's constantly regenerating. Every second, it orchestrates trillions of cellular operations without your conscious input. It speaks to you through sensations, cravings, and symptoms. But somewhere along the way, you were taught to ignore those messages. You were taught that the headache is a fluke, that the bloating is just "something you ate," that low energy is solved with caffeine, and that every irregularity should be silenced with a pill.

You were taught to normalize discomfort and to fear your body's own signals.

But what if those signals are wisdom? What if your fatigue is a call to rest and repair? What if your cravings are reflections of deep nutritional needs? What if your anxiety is a physiological response to hidden inflammation, gut imbalance, or toxic load?

This book doesn't offer magic bullets. It's not about chasing hacks or detox fads. It's about realignment—reconnecting with the design of your body and understanding the systemic factors that interfere with its innate healing power. The truth is: your body wants to heal. It knows how. It just needs the right inputs and fewer assaults.

The Blind Spots No One Told You About

Let's talk about what's been missing from the conversation. Not just in your doctor's office, but in the way we think about health on a societal level.

Why are some of the most disease-ridden nations also the most medicated?
Why do chronic illnesses continue to rise despite trillions spent on research and pharmaceutical development? Why are industrial seed oils, added sugars, and synthetic additives still allowed in foods labeled "healthy"? Why do medical schools gloss over the role of environmental toxins, circadian disruption, and microbiome balance?

The answers aren't simple, but the patterns are clear. Profit drives policy. Convenience trumps caution. And most of us have become passive recipients of narratives written by marketing departments, not scientists.

Even the language we use around health is flawed. We speak of "fighting disease" as if our bodies are battlegrounds, rather than nurturing systems trying to regain balance. We label symptoms as enemies instead of messengers. We fixate on what to eliminate—sugar, fat, carbs—without asking what to nourish.

Health is not a war. It is a relationship.

And like all relationships, it requires understanding, communication, and care.

Breaking Free from the Matrix of Misinformation

This book exists because many of us are waking up. We're asking better questions. We're not content with symptom suppression. We want root causes. We want resilience. We want to understand how to build lives that are not just free of disease, but full of vitality, clarity, and joy.

The journey begins with awareness. But awareness alone isn't enough. You also need tools. You need frameworks to evaluate the endless stream of health advice. You need ways to tune into your body's feedback and to filter out the noise.

This book gives you those tools.

It pulls back the curtain on the systems that shape your food, your medicine, your energy, your beliefs. It offers a new model of health—one that's grounded in biology, nature, and common sense. One that values prevention over intervention, nourishment over restriction, and connection over fragmentation.

You will explore:

- How your gut microbiome influences not just digestion, but mood, immunity, and mental clarity
- Why most people live in a state of hidden inflammation, and how to identify it before it erupts
- The quiet role of chronic stress, poor sleep, and blue light in derailing your hormones
- Why toxic exposures—chemical, emotional, and informational—accumulate and compound
- How ancient wisdom and cutting-edge science can work together, not in opposition
- What it means to truly take responsibility for your health in a culture that profits from your passivity

Each chapter is a doorway. A crack in the illusion. A path back to yourself.

A Call to Remember

This book is not an indictment of medicine or doctors. Many doctors are doing the best they can within a system that constrains them. But this book *is* a critique of the systems that shape what you're told, what's prescribed, and what's omitted.

And more than anything, it's a call to remember.

Remember that food is information. That your thoughts create chemistry. That movement is not punishment, but a celebration of life. That stillness heals. That nature is not optional—it's your original blueprint. That health is not a static destination—it is a rhythm, a dance, a daily act of listening.

You don't need to be perfect. You don't need to overhaul your life in one weekend. But you *do* need to wake up. Because no one is coming to save you—not the government, not the pharmaceutical industry, not even your doctor. The truth is, your body has been speaking to you all along. It's time to listen.

A New Chapter Begins

As you turn the page, I invite you to let go of what you think you know. Approach this journey not as a consumer of facts, but as a curious participant in your own healing. Question everything. Test it in your own experience. Be willing to unlearn as much as you learn.

What you'll find in these pages is not just information—it's *permission*. Permission to reclaim your sovereignty. To step outside the box of what's considered "normal" and build a version of health that feels alive, not just acceptable. To trust your intuition, your biology, your inner wisdom.

You've been lied to—not always maliciously, but consistently. And that ends now.

This is your invitation to step into truth. Not fear-based truth, not dogma—but grounded, empowering, embodied truth. The kind that doesn't come from institutions, but from your cells, your spirit, and your lived experience.

Welcome to the great remembering.

Welcome to *The Human Health Guide: The Missing Truths They Never Told You.*

Chapter 1: The Microbiome Revolution – Your Body's Hidden Universe

1.1 The Forgotten Organ: Understanding Your Microbial Ecosystem

Did you know that when you look in the mirror, only about 10% of what you're seeing is actually human? The rest—the invisible 90%—is made up of trillions of microbial hitchhikers. These aren't just passive residents. They are active, intelligent collaborators shaping almost every aspect of your health, from how you digest food to how you respond to stress. Welcome to the microbiome: a dense, dynamic ecosystem that has been quietly running the show from behind the scenes, often ignored by modern medicine until very recently.

For decades, scientists treated microbes as the enemy. Bacteria were to be eradicated. Fungi were seen as infections waiting to happen. Sterility was the gold standard, especially in hospitals, laboratories, and food production. But nature had other plans. As research accelerated in the early 2000s, something startling became clear: the bacteria we were trying to kill were often the very key to staying alive and well.

This realization gave birth to a new era—a revolution, really—in biology and medicine. It turned old models upside down. It challenged the notion of human autonomy and forced us to reconceptualize the body, not as a self-contained organism, but as a superorganism—a walking rainforest of life.

At the center of this revolution is the gut, an environment so rich in microbial diversity that it's now being referred to as the body's "forgotten organ."

The 100 Trillion Inhabitants Inside You

Your digestive tract, particularly your colon, houses more than 100 trillion microbial cells. That's more than all the stars in our galaxy. By comparison, your body contains only about 30 trillion human cells. In terms of genetic material, microbial DNA outnumbers human DNA in your body by a factor of over 100 to 1. This isn't just trivia—it's a radical shift in how we understand the human self.

These microbes don't just hang around doing nothing. They digest fibers that you can't break down. They manufacture essential nutrients like B vitamins and vitamin K. They train your immune system, regulate your metabolism, neutralize toxins, and keep harmful pathogens at bay. But perhaps most shockingly, they communicate directly with your brain, influencing your mood, your behavior, even your choices.

What makes this system so extraordinary is that it operates like an organ in its own right. Unlike your heart or liver, which are composed of human cells and remain relatively stable throughout life, your microbiome is fluid. It adapts. It evolves. It can be reshaped in weeks by what you eat, how you sleep, the drugs you take, and even the thoughts you think.

This microbial "organ" weighs between two and three pounds—roughly the same as your brain. That comparison isn't just poetic. The gut and the brain are intimately connected, and the gut microbiome is increasingly seen as a second brain, one that influences far more than digestion.

Microbial Fingerprinting: No Two Guts Alike

Just as no two fingerprints are the same, no two microbiomes are identical. Your gut's microbial profile is a signature—uniquely yours—formed from the moment you're born. In fact, how you entered the world played a major role in shaping it.

Were you born vaginally or via cesarean? Babies born through the birth canal are bathed in beneficial microbes from their mother's vaginal and intestinal flora. This early exposure seeds the infant's gut with foundational species that play a role in immune development and long-term health. Cesarean-born infants, on the other hand, are often colonized first by skin bacteria from the hospital environment—an initial imprint that may influence allergy, asthma, and metabolic risk later in life.

Feeding methods also leave their mark. Breast milk is not just food—it's an immunological language. It contains specific sugars, called human milk oligosaccharides, that aren't digestible by the baby but serve as a buffet for certain strains of beneficial bacteria like *Bifidobacterium infantis*. These microbes help seal the gut lining, reduce inflammation, and modulate the immune system.

Then there's antibiotic exposure. While antibiotics can be life-saving, they are also microbial napalm—wiping out good and bad bacteria alike. Studies have shown that even a single course of antibiotics can alter the gut microbiome for months, sometimes years. Repeated use, especially in early childhood, is linked to increased risks for obesity, autoimmune disorders, depression, and more.

Other shaping forces include diet (fiber-rich vs. processed), environment (urban vs. rural), household pets, stress levels, and even how much time you spend in nature. Each of these acts as a sculptor, carving the contours of your internal world.

What's astonishing is how little attention we give to this personal fingerprint in medicine. If someone had a liver that functioned radically differently from everyone else's, it would be the focus of every medical consult. But when it comes to our microbiomes—each as unique as our genome—we're still in the infancy of understanding.

Yet the implications are enormous. Two people can eat the exact same meal and experience wildly different metabolic responses based on their microbiota. This means that dietary advice, like medicine, must eventually become personalized. One-size-fits-all approaches are not just ineffective—they can be harmful.

The Gut-Brain Superhighway

Perhaps the most astonishing discovery of microbiome science is the intimate link between the gut and the brain. This relationship is not metaphorical—it's anatomical. A massive nerve called the vagus connects your gut directly to your central nervous system, acting as a bidirectional communication highway.

But the story doesn't stop with nerve signals. Your gut bacteria produce neurotransmitters—the very chemicals your brain uses to regulate mood and cognition. Serotonin, the molecule often targeted by antidepressants, is a prime example. Around 90% of your body's serotonin is synthesized in the gut, not the brain. Gut bacteria play a key role in regulating this process.

When your microbiome is imbalanced—a state known as dysbiosis—it can disrupt this entire communication loop. The result? Symptoms that don't appear "digestive" at all: anxiety, depression, irritability, brain fog, fatigue. It's no coincidence that people with irritable bowel syndrome (IBS) often experience comorbid mood disorders. What affects the gut affects the brain, and vice versa.

There's also mounting evidence that neurodegenerative diseases like Parkinson's may originate, in part, in the gut. Researchers have found that misfolded proteins implicated in Parkinson's appear in the gut years before motor symptoms emerge. This opens up the possibility that gut health might not only affect mental well-being but also neurological resilience over time.

Add to this the role of inflammation, which the gut microbiome also modulates. Chronic low-grade inflammation has been implicated in everything from Alzheimer's to schizophrenia. When gut bacteria are imbalanced, they can increase intestinal permeability—commonly referred to as "leaky gut." This allows bacterial components and undigested food particles to enter the bloodstream, triggering systemic inflammation and activating the immune system inappropriately.

What was once considered a fringe theory is now becoming mainstream. In 2022, the NIH launched a major initiative to explore the microbiome's role in mental health, a clear signal that the gut-brain axis is no longer pseudoscience—it's the future of medicine.

A Revolution in Perspective

Why does any of this matter? Because we are standing on the edge of a paradigm shift. For too long, health has been viewed through the lens of human cells, human organs, human DNA. But that view is incomplete. To understand ourselves fully, we must embrace the fact that we are ecosystems. We are not alone within our own skin.

This changes how we approach everything: food, medication, sanitation, stress, aging. It forces us to ask new questions. Not just "what does this do to me?" but "what does this do to my microbes?" When you eat a piece of processed food, you're not just feeding yourself—you're feeding your internal community. When you take an antibiotic, you're not just fighting an infection—you're wiping out thousands of microbial allies.

But the good news is this: the microbiome is resilient. It can be restored, nurtured, and diversified. Like a garden, it needs the right seeds, the right soil, and the right environment. And once it begins to thrive, you begin to thrive.

In the pages that follow, we'll dive deeper into how to rebuild this lost terrain—through nutrition, lifestyle, and environmental shifts. But for now, let this truth settle in: you are more than human. You are a living galaxy, a host to a microscopic universe that holds the keys to your health, your mood, and your future.

The revolution has already begun. The question is: are you ready to live in harmony with the hidden half of yourself?

1.2 The Microbiome-Disease Connection That Medicine Ignores

Despite the explosion of research around the microbiome in the past two decades, conventional medicine has been slow to integrate its insights into everyday clinical practice. There is still a large gap between what science now knows and what doctors actually apply in treatment rooms. As a result, countless patients continue to suffer from chronic conditions that are treated with prescriptions rather than understood at their roots. The truth is, many of the disorders that plague modern societies—from autoimmune diseases and allergies to depression and metabolic syndromes—are not mysterious afflictions that arise from nowhere. They are symptoms of an internal ecosystem in collapse.

One of the most underrecognized contributors to this collapse happens long before adulthood—often at birth. Cesarean deliveries, while sometimes medically necessary, are far more common than they need to be. In bypassing the birth canal, a child misses the first essential microbial bath: exposure to the mother's vaginal and fecal microbiota. Instead, they are colonized by the sterile environment of a hospital operating room—an initial microbial imprint that lacks the diversity and immune-regulating species nature intended. Add to this the fact that many of these newborns are given antibiotics in their first few days of life, often as a preventive measure, and you begin to see a pattern of disruption that can echo for decades.

Early-life antibiotic exposure is particularly problematic. The microbiome in infancy is not just a passive collection of bacteria; it is an architect of immune development. It trains the body to differentiate friend from foe, to tolerate harmless antigens while remaining vigilant against genuine threats. When this microbial education is cut short by antibiotics, the immune system becomes confused. It may begin to attack benign particles—like pollen, gluten, or even the body's own tissues. This is one reason why early antibiotic use is now associated with a heightened risk of autoimmune conditions, including type 1 diabetes, celiac disease, Crohn's, and multiple sclerosis, as well as allergic diseases like eczema and asthma.

This connection between microbial disruption and immune dysfunction is rarely discussed in the exam room. Instead, autoimmune diseases are treated as if they are unfortunate accidents—random misfirings of a faulty immune system, often addressed with immunosuppressants that silence symptoms without correcting the cause. But what if the immune system isn't broken at all? What if it's responding exactly as it was programmed to, by a microbiome that was shaped in dysbiosis from the very beginning?

This miscalibration continues throughout life. The widespread use of antibiotics for minor infections, the consumption of ultra-processed foods that starve beneficial bacteria, the over-sanitization of environments—all of these factors create the perfect storm for gut barrier breakdown. Known as "leaky gut," or more formally, increased intestinal permeability, this condition is at the heart of a modern epidemic of inflammation.

Under normal circumstances, your gut lining acts as a selective barrier, allowing nutrients to enter your bloodstream while keeping toxins, pathogens, and undigested food particles out. But when the microbiome is damaged—through poor diet, stress, toxins, or medications—the gut lining becomes compromised. Tight junctions between intestinal cells loosen, allowing substances that were never meant to circulate systemically to slip through the cracks.

This triggers an immune response. Your body recognizes these invaders as threats and mounts an attack, releasing inflammatory cytokines and activating white blood cells. Over time, this chronic, low-grade inflammation becomes systemic, affecting the brain, joints, skin, hormones, and virtually every organ system. It doesn't always cause digestive symptoms—many people with leaky gut experience no bloating or discomfort at all—but the downstream consequences are profound.

Depression, for instance, is increasingly understood as an inflammatory condition. Elevated markers like C-reactive protein and interleukin-6 are common in people with major depressive disorder. This has led researchers to investigate the "gut-immune-brain" axis, where microbial imbalances trigger gut permeability, which leads to inflammation, which then disrupts neurotransmitter function and mood regulation. It's a far more holistic and accurate model than the old "chemical imbalance" theory of serotonin deficiency.

Diabetes and obesity, too, are not just the result of overeating or laziness, as conventional wisdom once claimed. They are deeply influenced by the microbiome. Certain bacterial strains promote insulin sensitivity and energy balance, while others encourage fat storage and blood sugar spikes. When beneficial species are diminished—often by high-sugar, low-fiber diets—the metabolic system goes haywire. Insulin resistance takes root, fat accumulates, and hunger signals become distorted.

This is where the role of microbial metabolites becomes vital. Far from being passive byproducts, the substances produced by your gut bacteria are biochemically active. Short-chain fatty acids (SCFAs), such as butyrate, propionate, and acetate, are generated when fiber is fermented in the colon. These molecules reduce inflammation, strengthen the gut barrier, enhance insulin sensitivity, and even protect against cancer. They also influence the activity of regulatory T-cells, the peacekeepers of the immune system, which help prevent autoimmune flare-ups.

Bacteria also synthesize neurotransmitters like GABA, dopamine, and serotonin, all of which have powerful effects on mood, cognition, and stress resilience. In fact, some bacterial strains are now being researched as "psychobiotics"—organisms that can influence mental health through gut-mediated pathways. Vitamins like B12, folate, and K2 are likewise produced by a healthy microbiome, reinforcing the idea that your internal ecosystem isn't just an accessory to health—it is foundational to it.

Yet modern medicine often bypasses this connection entirely. When you present with depression, you're given a selective serotonin reuptake inhibitor. When you're diagnosed with an autoimmune disease, you're placed on corticosteroids or immune-modulating biologics. If you're pre-diabetic, you're put on metformin. In none of these scenarios is your microbiome tested, your diet examined, or your early microbial exposures considered. It is a blindness that costs lives.

But as with all revolutions, the tide is turning. Functional and integrative practitioners are now leading the way in connecting these dots, offering protocols that treat the microbiome as central, not peripheral. Patients are beginning to demand deeper answers. They are no longer satisfied with temporary relief. They want root cause resolution. And that resolution begins where few are taught to look—deep in the gut, in the lost kingdom of microbes that modern life has nearly forgotten.

1.3 Rebuilding Your Inner Ecosystem for Optimal Health

Once you understand how crucial your microbiome is to every aspect of your health, the next logical question is: how do I fix it? Unfortunately, the answer is not as simple as popping a probiotic pill or switching to Greek yogurt. While the supplement industry would have you believe otherwise, rebuilding a damaged microbial ecosystem requires intention, time, and an approach that goes far beyond surface-level fixes.

Let's start with probiotics. Most commercial probiotic supplements are composed of a few strains of *Lactobacillus* or *Bifidobacterium*, freeze-dried and encapsulated for convenience. While these can be helpful in certain scenarios—like recovering from a round of antibiotics—they often fail to colonize the gut long-term. The strains used are typically not human-native, and they pass through the digestive system like tourists rather than settlers.

More promising are spore-forming and soil-based organisms (SBOs), which have a natural resilience that allows them to survive the harsh acidity of the stomach and reach the colon intact. These microbes are more akin to the kinds our ancestors would have encountered regularly through contact with dirt, wild plants, and unchlorinated water sources. Today, we scrub our vegetables, sanitize our hands, and filter our environments to the point that these beneficial organisms are nearly extinct in urban populations. Reintroducing them—whether through specific SBO supplements or exposure to natural environments—is one of the most effective ways to diversify the gut.

Fermented foods are another potent tool. But not the pasteurized, vinegar-laden pickles and yogurts found on grocery shelves. We're talking about traditional, wild ferments: raw sauerkraut, kimchi, kefir, kombucha, miso, natto. These foods teem with live cultures, many of which are yet to be fully identified by science but have co-evolved with human digestion for thousands of years. Regular consumption of these foods can help reseed the gut with transient but therapeutically active microbes, while also providing enzymes and bioactive compounds that support digestion and immune function.

However, restoring the microbiome isn't just about adding bacteria. It's also about feeding the ones you already have. This is where prebiotics come in—non-digestible fibers that act as food for beneficial microbes. But not all prebiotics are created equal. Some, like resistant starch found in cooked-and-cooled potatoes or green bananas, selectively nourish butyrate-producing bacteria. Others, like polyphenols in berries and dark chocolate, serve a dual purpose: they modulate microbial populations while also acting as antioxidants and anti-inflammatories.

The challenge is that feeding the wrong types of bacteria can backfire. Those with SIBO (small intestinal bacterial overgrowth), for instance, often react poorly to high-FODMAP foods, which are rich in fermentable fibers. In these cases, timing and sequence matter. Sometimes, it's necessary to first reduce pathogenic overgrowth using targeted antimicrobial herbs like berberine, oregano oil, or allicin before rebuilding with prebiotics and ferments.

Then there's the frontier of microbiome restoration protocols. Fecal microbiota transplantation (FMT), once considered fringe and even repulsive, is now FDA-approved for treating recurrent *Clostridium difficile* infections and is being researched for everything from ulcerative colitis to Parkinson's. Though not widely available, its results hint at a profound principle: sometimes the best way to reset a damaged microbiome is to borrow a healthy one.

For those not ready to undergo FMT, there are gentler but still powerful options. Rotating strains of probiotics to mimic microbial diversity, using binders to clear out endotoxins released by dying pathogens, and consuming mucosal-supportive compounds like L-glutamine and colostrum can all aid in restoring gut integrity. Precision testing—such as stool analysis, breath tests, and metabolomics—can guide these interventions with increasing specificity, tailoring treatment to your unique microbial profile.

Ultimately, restoring your microbiome is not a one-time protocol. It is a lifestyle. It involves moving your body regularly to stimulate peristalsis and lymphatic flow. It means reducing exposure to antibiotics, NSAIDs, and chemicals like glyphosate that harm gut flora. It involves managing stress, which profoundly alters gut function via the hypothalamic-pituitary-adrenal (HPA) axis. And it calls for reconnection—with the earth, with living foods, with ancient practices that saw the human body not as a machine, but as a garden to be tended.

In reclaiming your microbiome, you are not just fixing a digestive issue. You are rebuilding the foundation for immunity, mental clarity, emotional resilience, and metabolic vitality. You are participating in the restoration of something that modernity tried to sterilize out of existence—the intimate, intelligent, living web of life within you.

And in doing so, you don't just heal your own body. You help to restore a wisdom that our ancestors always knew but that science is only now beginning to rediscover: health is not a solitary achievement. It is a symphony of life, inside and out.

Chapter 2: Circadian Biology – The Master Clock They Forgot to Tell You About

2.1 The Circadian Command Center: How Light Controls Your Life

Picture this: an astronaut floating weightless in the vast stillness of space. No sunrise. No sunset. No natural day or night. It might sound like the pinnacle of freedom—untethered from Earth's rhythms—but what NASA discovered was far from liberating. Deprived of normal light-dark cycles, astronauts' biological systems began to unravel. Cellular aging accelerated. Hormonal rhythms collapsed. Immune function weakened. It wasn't just the absence of gravity or the exposure to radiation that caused these problems. It was the disruption of time itself—biological time.

Back on Earth, we are replicating the same dysfunction, but on a slower, more invisible scale. We've constructed a 24/7 society where the lights never go off, the screens never dim, and the rhythm of the planet has been replaced by the rhythm of productivity. We eat at midnight. We binge-watch under blue LEDs. We wake to alarms rather than sunlight. And then we wonder why we're tired, overweight, anxious, and sick. The truth is, we've forgotten we are rhythmic creatures. And at the core of this rhythm lies a biological clock so powerful, so all-encompassing, that it orchestrates nearly every function in the body.

Welcome to the world of circadian biology—a field that is quietly revolutionizing our understanding of health, performance, aging, and disease.

At the heart of this revolution is a microscopic region of the brain called the **suprachiasmatic nucleus** (SCN). Tucked within the hypothalamus, just above the optic chiasm where the optic nerves cross, the SCN contains approximately 20,000 neurons. That's it. Twenty thousand tiny nerve cells in a sea of billions—and yet they serve as the master conductor of your body's entire symphony. Every hormone pulse, every metabolic switch, every digestive enzyme, every cellular repair mechanism—all of it is timed, and the conductor's baton is light.

From the moment photons of morning sunlight hit your eyes, the SCN springs into action. It interprets this light as a signal: the day has begun. The SCN then relays this message to other parts of the brain and the body, initiating cascades of biochemical events. Cortisol begins to rise, helping you feel alert and mobilized. Melatonin production is shut off. Body temperature starts to climb. Digestive organs prepare for feeding. Your physiology is not passively existing—it is preparing, predicting, anticipating, and aligning itself with the natural cycle of day and night.

This rhythm—roughly 24 hours in length—is what scientists call a *circadian* rhythm, from the Latin *circa diem*, meaning "about a day." And while the existence of such cycles has been known for centuries—farmers observed it long before biologists named it—it wasn't until relatively recently that we began to understand the depth of its influence. It's not just sleep and wakefulness that follow a circadian pattern. Virtually every system in the body—immune, cardiovascular, digestive, hormonal, neurological—is under circadian regulation.

The SCN acts as the master clock, but it doesn't work alone. Every cell in your body contains its own molecular clock, a tiny gearwork of genes and proteins that cycle on a 24-hour rhythm. These are known as **peripheral clocks**, and they are found in the liver, heart, kidneys, pancreas, lungs—even in your skin. They're incredibly precise, but also incredibly dependent on timing cues—called *zeitgebers*—to stay in sync. The strongest zeitgeber is light.

For decades, we believed that the eyes were simply windows to the brain—taking in light to form images. But that changed with the discovery of a remarkable type of cell: the **intrinsically photosensitive retinal ganglion cell**, or ipRGC. Unlike rods and cones, which help us see images and motion, ipRGCs are not involved in vision at all. They don't form pictures. They don't respond to color or shape. Their sole purpose is to detect **blue light**—specifically, light in the wavelength of around 460–480 nanometers—and to send that information directly to the SCN.

This was a revelation. It meant the body had a fourth photoreceptor system, entirely separate from vision, devoted solely to the regulation of time. These melanopsin-containing cells function like biological light meters, measuring the brightness of the environment and helping calibrate the master clock accordingly. When you step outside into natural sunlight—even on a cloudy day—these cells are flooded with blue light, and your SCN responds with clarity: it is daytime.

This interaction is not subtle. Even a few minutes of bright morning light can reset your circadian clock and anchor it for the day. This is why exposure to natural light in the early hours is one of the most powerful health interventions available—completely free, yet rarely practiced. And it's also why the modern habit of checking phones in bed, sitting under LED lights late into the night, and working in artificially lit cubicles during the day is so devastating to health.

The problem isn't just light at night—it's *the wrong kind* of light at *the wrong time*. Artificial lighting, particularly the blue-heavy spectrum of screens and LED bulbs, hijacks the melanopsin pathway. It tricks the SCN into believing it's still daytime, suppressing melatonin production and delaying the body's transition into night mode. Melatonin isn't just a sleep hormone—it's a master regulator of cellular repair, immune function, and antioxidant defense. Suppress it consistently, and you're suppressing regeneration at the deepest levels.

But the cascade doesn't stop there. Once the master clock is confused, it falls out of sync with the peripheral clocks. The liver, for instance, anticipates feeding times and gears up to metabolize nutrients. If you eat late at night, when your liver is winding down, digestion becomes inefficient, glucose regulation is impaired, and fat storage increases. Likewise, the pancreas has its own clock that regulates insulin secretion. If its rhythm is misaligned with your actual eating schedule, insulin resistance can develop, paving the way for metabolic disorders.

The kidneys follow daily patterns of fluid balance and blood pressure regulation. Disrupt their timing, and you get irregularities in electrolyte balance and hypertension. The heart, too, has a circadian rhythm—so precise that most heart attacks occur between 6:00 and 11:00 a.m., when the sympathetic nervous system naturally peaks. Even the immune system has time-dependent behaviors, launching attacks on pathogens more effectively at certain times of day.

All of this adds up to a profound realization: your body is not static. It is rhythmic. Every process has a right time and a wrong time. And when those rhythms are respected, you thrive. When they are ignored, you unravel—slowly at first, then all at once.

This helps explain a growing body of research linking circadian disruption to nearly every chronic illness of modern life. Shift workers, who live in perpetual circadian misalignment, suffer higher rates of obesity, cancer, diabetes, heart disease, and depression. Night owls forced to wake early experience cognitive impairment and immune suppression. Even short-term disruptions—like jet lag or daylight saving time—can increase accident risk, destabilize mood, and throw off metabolic function.

And yet, how often do doctors ask about your light exposure, your sleep timing, your meal schedule? How often are chronic symptoms viewed through the lens of circadian rhythm? Almost never. It is a blind spot in modern medicine, one with enormous consequences.

But there is hope in this, too. Because circadian rhythms are *malleable*. They are programmable. Your body is constantly listening to environmental cues, ready to realign if you give it the right signals. By anchoring your day with light in the morning, darkness in the evening, consistent meal times, and regular sleep patterns, you can restore harmony to your internal clocks. You can reclaim energy, clarity, and resilience—not through pills or hacks, but by syncing your life with the natural order.

What NASA learned in space, we must remember on Earth: time is not just measured by clocks. It is lived through light. Through darkness. Through rhythm. You don't need to understand the biochemistry of PER and CLOCK genes, or memorize every enzyme tied to circadian cycles. You simply need to respect the fundamental truth that your biology was designed for a planet that turns. When you move with it, everything begins to align. When you fight it, everything begins to fray.

In the chapters that follow, we'll explore how to live in step with your master clock—not by withdrawing from modern life, but by reshaping your environment to work *for* your biology, not against it. For now, remember this: your health is not just about what you eat, how much you move, or even how you think. It is also about *when*. Timing is not a detail—it is the architecture of life itself.

2.2 The Hidden Health Costs of Circadian Disruption

Modern life has quietly engineered a massive experiment—one in which billions of people live out of sync with their biological clocks, often without knowing it. We've redefined productivity, normalized late nights, glorified hustle culture, and surrendered our internal rhythms to artificial schedules, artificial lights, and artificial meals. The consequences are staggering, and while they may appear subtle at first—difficulty falling asleep, low energy, mild anxiety—the toll they take over months and years is anything but minor. This is not just about being a "morning person" or a "night owl." Circadian disruption is now recognized as a foundational driver of many of the chronic diseases that define the 21st century.

One of the most extreme examples of circadian misalignment can be seen in night shift workers. These are the nurses, factory workers, airline staff, security personnel, and countless others who live in direct contradiction to their internal clocks. The term "shift work disorder" has been coined to describe the cluster of symptoms they commonly experience: insomnia, excessive daytime sleepiness, impaired concentration, and mood disturbances. But beneath the surface, something even more troubling is happening.

When people work through the night and sleep during the day, their melatonin production is suppressed by exposure to artificial light, while their cortisol levels remain elevated due to the stress of being awake at biologically inappropriate times. This hormonal confusion leads to a cascade of disruptions. Over time, night shift workers exhibit up to a 40% higher risk of developing cardiovascular disease, type 2 diabetes, certain cancers (notably breast and prostate), and neurodegenerative disorders. The International Agency for Research on Cancer (IARC), a branch of the World Health Organization, has even classified night shift work as a probable carcinogen.

The accelerated aging seen in these individuals is not an exaggeration. Cellular repair mechanisms, which are largely governed by circadian rhythms, falter when the internal clock is thrown off course. Telomere shortening, mitochondrial dysfunction, and increased oxidative stress—all hallmarks of biological aging—become more pronounced. In essence, those who live out of rhythm age faster, not because they are working harder, but because they are working against time itself.

But you don't have to work nights to suffer the consequences of circadian disruption. Enter the phenomenon of "social jet lag." This term describes the growing discrepancy between our biological clocks and our social clocks—particularly the difference between our sleep schedules during the workweek and on weekends. For millions of people, Monday through Friday is dictated by alarm clocks, early commutes, and regimented wake times. But when Saturday arrives, they sleep in, stay up late, and allow their natural preferences to surface. This temporary shift, often only an hour or two in either direction, may feel like a relief. But biologically, it's a mini jet lag every week—and the effects are cumulative.

Studies have shown that even a single hour of circadian misalignment can increase the risk of metabolic disorders. One prominent study found that just one hour of social jet lag raised the risk of type 2 diabetes by 27%. Why? Because the body's insulin sensitivity is tightly bound to circadian rhythms. When meal timing is inconsistent and sleep is erratic, the hormonal orchestration that governs appetite, fat storage, and glucose regulation becomes chaotic. The result is not just weight gain, but a systemic weakening of metabolic control. Mood swings, cravings, and even immune system vulnerability follow closely behind.

And while social jet lag may appear trivial—after all, who hasn't slept in on a Sunday?—the regular disruption of time cues leaves the body confused. It's the biological equivalent of trying to dance to a song when the beat keeps changing. The rhythm falters. Coordination breaks down. And eventually, the entire system loses its timing.

Then there is the most pervasive and insidious culprit of all: **light pollution**. This is not the dramatic kind we think of when gazing at a city skyline from above. It's the soft glow of your phone screen in bed. The overhead LEDs in your kitchen. The streetlights seeping through your curtains. Exposure to artificial light after sunset, especially blue light, hijacks your brain's perception of time and delays the natural cascade of nighttime physiology.

The effects on metabolism are significant. Leptin, the hormone that signals satiety and tells the body to stop eating, is sensitive to circadian timing. Disrupted light exposure decreases leptin sensitivity, which leads to overeating and altered feeding behaviors. This is one reason people tend to eat more at night—and why those calories are more likely to be stored as fat. Simultaneously, insulin resistance increases, making it harder for cells to absorb glucose and further contributing to weight gain and metabolic dysfunction.

Obesity, in this context, is not simply a matter of willpower or calories in versus calories out. It is often a matter of timing. When the body is flooded with food at a time it was designed to rest and repair, metabolic chaos ensues. Artificial light extends the day artificially, delaying melatonin and promoting late-night eating. The feedback loop is vicious: poor sleep increases cravings, circadian disruption increases fat storage, and disrupted feeding times impair insulin action.

What's particularly disturbing is how normalized this has become. Late-night Netflix, 24-hour food delivery, scrolling endlessly through social media before bed—all of these behaviors are now default. But just because they are common does not mean they are harmless. The body still remembers how it was designed. And it reacts accordingly. With inflammation. With fatigue. With disease.

As we move deeper into a world governed by screens, algorithms, and artificial rhythms, the need to reclaim our biological timing becomes urgent. The health costs of ignoring circadian biology are no longer speculative—they are real, measurable, and widespread. But the same system that is so easily thrown off track can also be brought back into balance—if we understand how to work with it rather than against it.

2.3 Mastering Your Circadian Rhythm for Peak Performance

If disruption of the circadian rhythm lies at the root of so many modern ailments, then restoration of that rhythm holds immense potential—not just for preventing disease, but for enhancing energy, cognition, metabolism, and longevity. The circadian system is a tool, a rhythm, a master key to biological coherence. The problem is, most people don't know how to wield it. But once you understand how to engage its natural gears—through light, food, and temperature—you unlock a level of performance and resilience that pharmaceutical solutions rarely deliver.

The cornerstone of circadian alignment is light. Not just any light, but **strategically timed light exposure**. The goal is simple: tell your body, with absolute clarity, when it is day and when it is night. The most powerful way to do this is by stepping outside within 30 to 60 minutes of waking and getting real sunlight into your eyes—without sunglasses, and ideally for at least 20 to 30 minutes. This exposure calibrates the SCN and triggers a cascade of wakeful hormones and alertness chemistry. It tells the body: "The day has begun." The stronger this signal, the stronger your circadian rhythm will be.

For those who live in northern latitudes or face dark winters, **light therapy boxes** can be a powerful substitute. Devices that emit 10,000 lux of full-spectrum light used early in the morning can provide enough stimulation to regulate mood, energy, and hormonal timing. These tools have even been shown to be as effective as antidepressants for seasonal affective disorder and are increasingly used by athletes, shift workers, and biohackers to reinforce morning wakefulness.

Equally important is the absence of light in the evening. After sunset, the goal is to reduce exposure to blue light. This means dimming overhead lights, switching to amber bulbs, and using blue light blocking glasses if screen use is unavoidable. Even better, switch to **red-spectrum lighting** in your living space, which does not interfere with melatonin production. Unlike white or blue light, red light supports the body's transition into rest without disrupting circadian chemistry. In fact, red light therapy, when applied strategically in the morning or afternoon, can enhance mitochondrial function and support cellular energy production through increased ATP synthesis. It does this without suppressing melatonin, making it a powerful tool for performance and recovery.

Food timing is the second pillar of circadian mastery. More than what you eat, *when* you eat shapes your metabolic outcomes. This is the principle behind **circadian fasting** or **time-restricted eating**—limiting food intake to an 8–12 hour window during the day, ideally ending before sunset. Studies have shown that eating in alignment with daylight hours improves glucose regulation, reduces inflammation, enhances fat oxidation, and activates

autophagy—the cellular cleanup process that removes damaged proteins and renews cellular components.

The reason circadian fasting works so well is because the body expects food at certain times. Enzymes, bile acids, insulin sensitivity—all follow predictable daily rhythms. Late-night eating, even if it's healthy food, places a burden on a system that has already begun shifting into rest and repair. And over time, this misalignment wears down metabolic efficiency.

The third powerful circadian cue is **temperature**. Body temperature follows a natural rhythm, peaking in the late afternoon and dropping during the night. This drop is a signal for sleep initiation. You can amplify this signal by manipulating your environment. Cold exposure—whether through cold showers, ice baths, or cryotherapy—can reinforce alertness in the morning and recalibrate the nervous system. In contrast, sauna use in the evening, followed by a cool-down, can deepen sleep by accelerating the natural decline in core temperature.

Bedroom cooling is perhaps the most underrated performance enhancer. Keeping your sleeping environment cool—ideally between 60 to 67°F (15 to 19°C)—supports deep, restorative sleep by aiding the thermoregulatory decline that signals the brain it's time to release melatonin and enter non-REM stages. Even slight overheating can fragment sleep cycles, reduce slow-wave sleep, and impair overnight recovery.

What ties all of this together is consistency. The circadian system thrives on rhythm. One day of perfect alignment followed by two days of chaos won't cut it. What matters is the regularity of the signals—light, food, movement, and rest. It's not about perfection, but about predictability. Your body doesn't need rigid control. It needs rhythm, and rhythm requires repetition.

By becoming intentional with your exposure to light, the timing of your meals, and the manipulation of your internal temperature, you're not just managing sleep—you're optimizing an entire hormonal symphony. Testosterone, estrogen, growth hormone, insulin, thyroid hormones—all of them pulse on daily cycles. When you align with your circadian rhythm, you ride the waves of these natural fluctuations instead of crashing against them.

Circadian mastery is not a wellness trend. It's the foundation of human biology. Before nutrition, before fitness, before mindfulness—there was rhythm. It is not optional. It is the architecture of life. And the more we return to it, the more health stops being a struggle and becomes a consequence of living in sync with who we truly are.

Chapter 3: The Mitochondrial Energy Crisis – Why You're Always Tired

3.1 Mitochondria: The Cellular Powerhouses Under Attack

You may not have heard of them in high school beyond a passing mention in biology class, but your mitochondria are the unsung heroes of your health. These microscopic structures inside your cells are responsible for keeping you alive at the most fundamental level—not just functioning, but thriving. Without them, you wouldn't breathe, move, think, or exist. Yet despite their vital importance, they are under constant assault in the modern world. When they falter, you feel the consequences not just as fatigue, but as full-body burnout: metabolic disorders, brain fog, mood disturbances, premature aging, and chronic illness. The energy crisis you feel in your life—whether it's mental exhaustion, physical sluggishness, or emotional flatness—begins at the mitochondrial level.

To understand why, we need to take a step back. Way back. About 1.5 billion years ago, to be precise. Long before the human body ever existed, two ancient life forms entered into a partnership that would become the blueprint for all complex life: a primitive eukaryotic cell engulfed a small, energy-producing bacterium. But instead of digesting it, something unprecedented happened. The two organisms began to cooperate. The engulfed bacterium, which could convert oxygen and nutrients into vast amounts of energy, stayed on board. Over millions of years, it lost some of its independence but retained its own DNA. It specialized in energy production, while the host cell provided protection and raw materials.

This ancient alliance is still alive inside you today. Every cell in your body (with a few exceptions, like red blood cells) contains dozens, hundreds, or even thousands of mitochondria. These aren't just passive machines. They are descendants of those early bacteria—semi-autonomous organisms with their own genome, their own replication processes, and their own vulnerabilities. And they are still tasked with the same mission: turning the food you eat and the oxygen you breathe into usable energy.

That energy comes in the form of adenosine triphosphate, or ATP. ATP is the universal energy currency of biology. It powers every heartbeat, every nerve impulse, every breath, every movement, every thought. A single human body produces its own weight in ATP every day. That's how essential this molecule is. And mitochondria are the factories where it's made.

The process by which mitochondria produce ATP is called **oxidative phosphorylation**, and it happens in a beautifully complex series of steps known as the **electron transport chain**. This chain is composed of four main protein complexes (I through IV) embedded in the inner mitochondrial membrane. Electrons, stripped from the food you eat, are passed along these complexes in a highly controlled sequence. As they move through the chain, they release energy, which is used to pump protons across the membrane and create an electrochemical gradient. This gradient is then harnessed by a fifth complex— ATP synthase—which acts like a molecular turbine, churning out ATP with astonishing efficiency.

When this system is working well, the cell runs like a well-oiled machine. Energy is abundant, cellular repair is brisk, detoxification is effective, and biological processes operate at peak performance. But when even one step of this chain is disrupted—by toxins, inflammation, oxidative stress, nutrient deficiency, or genetic mutation—energy production can collapse. And because the process is exponential, not linear, small disruptions can have massive effects. Imagine a factory where a single broken cog brings the entire assembly line to a halt. That's what happens when mitochondrial function is impaired.

The consequences ripple outward. Cells begin to function less efficiently. Some die prematurely. Others limp along, barely able to perform their tasks. The tissues that rely most heavily on mitochondria—like the brain, heart, muscles, and liver—are the first to suffer. Fatigue sets in. Memory dulls. Muscles ache. Hormones fall out of rhythm. The immune system weakens. You may not feel "sick" in a way a doctor can diagnose, but you no longer feel well. You've lost access to your full vitality.

This is why mitochondrial health is central to almost every chronic condition. In neurological disorders like Alzheimer's and Parkinson's, damaged mitochondria accumulate in brain cells, leading to impaired synaptic function and neuronal death. In cardiovascular disease, mitochondrial dysfunction compromises the energy supply to heart tissue, making it vulnerable to stress and inflammation. In metabolic conditions like diabetes and obesity, inefficient mitochondria reduce the cell's ability to burn fat and sugar, creating energy imbalances and increasing oxidative damage. Even cancer has a mitochondrial component, as many tumor cells exhibit abnormal energy metabolism, favoring inefficient pathways like glycolysis even in the presence of oxygen—a phenomenon known as the Warburg effect.

But why are mitochondria so vulnerable? The answer lies partly in their evolutionary history. Because they originated as bacteria, mitochondria retained their own DNA, which is separate from the DNA housed in your cell nucleus. This mitochondrial DNA (mtDNA) is small, circular, and encodes just 37 genes—most of which are essential for energy production. But unlike nuclear DNA, which is protected by histones and shielded within the nucleus, mtDNA floats freely within the mitochondrial matrix. This makes it particularly susceptible to damage from free radicals, which are produced as natural byproducts of energy metabolism.

Under normal conditions, mitochondria have antioxidant defenses to neutralize these radicals. But when those defenses are overwhelmed—by environmental toxins, chronic stress, poor diet, infection, or sleep deprivation—free radicals attack the mtDNA, causing mutations and impairing energy production even further. It's a vicious cycle: damaged mitochondria produce more free radicals, which cause more damage, leading to a downward spiral of cellular aging and dysfunction.

To make matters worse, mitochondrial DNA has limited repair capacity. It lacks the sophisticated error-correction mechanisms that nuclear DNA enjoys. Once damaged, mitochondrial genomes are difficult to fix. Over time, these mutations accumulate, leading to what scientists call "mitochondrial heteroplasmy"—a mix of healthy and damaged mitochondria within the same cell. As the proportion of damaged mitochondria grows, the cell's overall function declines. This is one of the key mechanisms behind the aging process.

But mitochondrial damage doesn't just come from within. The modern world is filled with external threats to mitochondrial health. Environmental toxins like heavy metals (mercury, lead, cadmium), pesticides (glyphosate, organophosphates), and air pollutants directly impair mitochondrial enzymes. Certain medications, such as statins, antibiotics, and chemotherapy agents, can also damage mitochondrial membranes or disrupt the electron transport chain. Even electromagnetic fields (EMFs) from wireless devices are now being investigated for their potential to interfere with mitochondrial signaling and membrane integrity.

Then there's the impact of lifestyle. Sedentary behavior reduces the demand for mitochondrial energy production, leading to atrophy. Poor sleep disrupts the circadian regulation of mitochondrial biogenesis—the process by which new mitochondria are created. Chronic psychological stress floods the body with cortisol, which in excess impairs mitochondrial gene expression and lowers antioxidant capacity. High-sugar diets overload the system with glucose, forcing mitochondria to burn fuel inefficiently and increasing oxidative stress. It's no wonder fatigue is the most common complaint in modern clinical practice.

Despite all this, mitochondria are not fragile relics of ancient life. They are remarkably adaptive—if given the right inputs. They can increase in number, grow in size, and improve their efficiency in response to signals like exercise, caloric restriction, cold exposure, and certain phytonutrients. But they need time, space, and nourishment to recover. They need oxygen. They need rest. They need micronutrients like magnesium, CoQ10, carnitine, B vitamins, and alpha-lipoic acid to function properly. They need a break from the relentless onslaught of synthetic chemicals and metabolic overload.

Most of all, they need to be remembered. For too long, mitochondria have been relegated to the background of medicine, overshadowed by genes, hormones, and organs. But they are not background characters. They are the engine. They are the flame. And when that flame is low, nothing burns properly.

Rekindling that flame requires a new perspective on health—one that moves beyond treating symptoms and starts restoring energy at the cellular level. In the next section, we will explore how to support mitochondrial resilience through targeted lifestyle interventions, natural therapies, and emerging scientific tools. But for now, remember this: your energy is not a mystery. It's not laziness. It's not aging. It's mitochondrial. And when you protect your mitochondria, you protect your life force.

3.2 Modern Mitochondrial Toxins and Energy Saboteurs

To understand why so many people today feel chronically tired, foggy-headed, or inexplicably inflamed, we have to move beyond vague notions of "stress" or "aging" and look deeper—into the toxic landscape in which our mitochondria are now forced to operate. The truth is uncomfortable but undeniable: we are bathing in a chemical soup that did not exist a hundred years ago. Our air, water, soil, food, and even the devices we use daily are sources of chronic mitochondrial interference. This isn't fearmongering—it's biochemistry. And when you grasp the degree to which environmental toxins and manmade stressors affect your mitochondria, fatigue stops being a mystery.

Let's begin with heavy metals. These aren't some abstract industrial concern—they're inside us. Mercury, lead, cadmium, and arsenic have no biological function in the human body, and yet they accumulate relentlessly in tissues, particularly in the nervous system and mitochondrial membranes. Mercury, for example, has an affinity for sulfhydryl groups, which are present in mitochondrial enzymes essential for energy production. Once mercury binds, it doesn't simply pass through the body; it disrupts enzyme activity, increases oxidative stress, and impairs the flow of electrons through the electron transport chain.

Cadmium, commonly found in cigarette smoke, industrial emissions, and some cheap jewelry or paints, also interferes with mitochondrial respiration. It displaces zinc and selenium—two key minerals required for antioxidant enzymes like superoxide dismutase and glutathione peroxidase. This displacement creates a condition where oxidative stress skyrockets and mitochondria are unable to defend themselves. Over time, mitochondrial membranes begin to break down, and ATP production drops significantly.

Pesticides such as organophosphates and glyphosate don't just affect plants and pests. They also impact mammalian biology, particularly mitochondrial bioenergetics. Some of these chemicals act as direct inhibitors of cytochrome enzymes in the electron transport chain. Others generate excess reactive oxygen species (ROS), overwhelming the body's antioxidant defenses. Long-term exposure—even at low doses—can lead to mitochondrial swelling, membrane leakage, and reduced oxygen consumption at the cellular level.

The body has limited mechanisms to detoxify these substances once they enter the mitochondria. Unlike the nucleus, mitochondria lack robust repair systems. Damage tends to accumulate, and as it does, cells begin to lose their resilience. Muscles fatigue more easily. The brain loses its sharpness. The immune system becomes sluggish or overreactive. And because mitochondria are involved in cellular signaling and apoptosis (programmed cell death), their dysfunction often becomes the root of broader systemic breakdown.

But the assault on mitochondrial integrity doesn't stop with environmental exposure. **Pharmaceuticals**, often trusted as life-saving tools, can also act as mitochondrial saboteurs when misused or overprescribed. Statins, widely used to lower cholesterol, also block the body's synthesis of Coenzyme Q10—a vital component of the electron transport chain. Without enough CoQ10, electrons cannot be transferred efficiently, and ATP production stalls. This explains why many statin users complain of muscle pain, weakness, and fatigue. It's not a rare side effect—it's a biochemical inevitability in the context of depleted mitochondrial support.

Antibiotics, especially broad-spectrum varieties, do more than wipe out gut bacteria. Some, like fluoroquinolones, have been shown to damage mitochondrial DNA directly. Others inhibit mitochondrial protein synthesis, leading to compromised energy output in cells that rely heavily on constant energy, such as neurons and muscle fibers. What's more, the gut microbiome and mitochondria are closely linked. Gut dysbiosis, often caused by antibiotics, can create an inflammatory state that indirectly impairs mitochondrial performance system-wide.

Psychiatric medications—including certain antipsychotics, SSRIs, and benzodiazepines—also have mitochondrial consequences. Many interfere with mitochondrial membrane potential, calcium regulation, or complex I activity. In some cases, prolonged use has been associated with increased oxidative stress and reduced neuroplasticity, potentially worsening the very cognitive symptoms they are meant to treat.

Unfortunately, few physicians are trained to recognize these effects as mitochondrial in origin. Patients who present with fatigue, brain fog, or pain are often told they're experiencing stress, aging, or depression. The deeper cause—mitochondrial suppression due to chemical or pharmaceutical exposure—is rarely considered. This blind spot leads to further prescriptions, layering mitochondrial toxicity without addressing its root.

Then comes the invisible layer of exposure: **electromagnetic fields (EMFs)**. Wireless technology, for all its conveniences, operates within frequencies that are biologically active. Mitochondria are electrochemical engines, meaning they are inherently sensitive to electromagnetic inputs. Emerging research has shown that chronic exposure to EMFs from Wi-Fi, smartphones, Bluetooth, and other devices can disrupt mitochondrial calcium signaling—a crucial process for ATP synthesis, cell communication, and immune regulation.

When calcium channels on the mitochondrial membrane are dysregulated, excess calcium floods into the mitochondria, leading to increased ROS generation and impaired oxidative phosphorylation. This can result in reduced energy output, increased cellular apoptosis, and inflammation. Some individuals are more sensitive than others, experiencing symptoms like headaches, insomnia, palpitations, and chronic fatigue in the presence of strong EMF sources. This condition, known as electromagnetic hypersensitivity (EHS), is still controversial in mainstream medicine, but its physiological basis is increasingly supported by mitochondrial studies.

The term "dirty electricity" refers to erratic electrical frequencies emitted from poorly filtered power supplies or electrical devices. These surges can introduce chaotic patterns into the body's electrical systems, disrupting cellular rhythms, including those of the mitochondria. While the exact mechanisms remain under investigation, clinical observations repeatedly show improvement in energy, sleep, and mood when patients reduce EMF exposure and improve their living environments with grounding practices and shielding.

The combined effects of these toxic exposures are cumulative. No single toxin may be the tipping point, but together, they create a silent collapse of mitochondrial integrity. The modern world is not just energy-draining metaphorically—it is energy-draining biologically. And until we recognize the ways we're sabotaging our mitochondria, we will continue to chase symptoms, misunderstanding the true source of our decline.

3.3 Mitochondrial Restoration and Biogenesis Protocols

While it may seem overwhelming to face the constant barrage of mitochondrial insults in daily life, the good news is that mitochondria are not passive victims. They are highly responsive to the right stimuli—if you provide the proper conditions, they don't just recover; they multiply, adapt, and upgrade. The process of creating new, healthy mitochondria is known as **mitochondrial biogenesis**, and activating it is one of the most powerful ways to rebuild your energy and resilience from the ground up.

One of the most promising tools in this regard is the synergy between **PQQ (pyrroloquinoline quinone)** and **Coenzyme Q10 (CoQ10)**. PQQ is a redox cofactor found in soil and certain plant-based foods, but in supplement form, it has been shown to stimulate the expression of genes involved in mitochondrial biogenesis. In essence, it tells your cells to start building more power plants. CoQ10, on the other hand, is not involved in building new mitochondria but is critical for helping them function. It's embedded in the inner mitochondrial membrane, where it shuttles electrons between complexes I and III of the electron transport chain, ensuring a smooth and efficient flow of energy.

Together, PQQ and CoQ10 form a complementary pair: one promotes the creation of new mitochondria; the other ensures they operate at full efficiency. Clinical doses often range from 10 to 20 mg daily of PQQ and 100 to 300 mg daily of ubiquinol (the active form of CoQ10), taken with fat-containing meals to enhance absorption. Many individuals report improved mental clarity, stamina, and sleep within weeks of starting this combination.

But supplementation alone isn't enough. The body's most powerful signal for mitochondrial regeneration is **hormetic stress**—short, controlled doses of physiological challenge that trigger adaptation. **Cold thermogenesis**, or deliberate exposure to cold temperatures, is one of the most effective ways to initiate this process. When you immerse yourself in cold water or take a cold shower, your brown adipose tissue (BAT) becomes activated. BAT is rich in mitochondria and burns energy to generate heat. This not only increases mitochondrial density in

BAT itself but sends systemic signals to enhance mitochondrial efficiency throughout the body.

Cold exposure also increases norepinephrine, a neurotransmitter that promotes focus and arousal, and boosts the expression of uncoupling proteins (UCPs), which improve metabolic flexibility. Over time, regular cold exposure can retrain your mitochondria to produce more energy with less waste, increasing your endurance and recovery. Start with 30 seconds at the end of your normal shower and gradually work up to full cold showers or ice baths. The key is consistency and gradual adaptation.

Nutritional support is equally critical. Mitochondria are fueled not just by macronutrients but by a host of micronutrients that act as cofactors in energy metabolism. **NAD+ precursors**, such as NMN (nicotinamide mononucleotide) or NR (nicotinamide riboside), have gained attention for their role in mitochondrial repair and gene regulation. NAD+ is required for the activity of sirtuins, a family of proteins involved in cellular longevity and mitochondrial biogenesis. Supplementing with 250–500 mg of NMN or NR daily may support mitochondrial renewal, especially when paired with intermittent fasting or exercise, both of which naturally raise NAD+ levels.

Alpha-lipoic acid (ALA) is another potent mitochondrial antioxidant that helps regenerate other antioxidants like glutathione and CoQ10. It also improves insulin sensitivity and assists in glucose metabolism, making it ideal for those with fatigue rooted in blood sugar instability. Typical doses range from 300 to 600 mg per day, often divided into morning and early afternoon doses.

Creatine, often thought of as a bodybuilder's supplement, is actually a powerful mitochondrial supporter. It helps buffer cellular energy by donating phosphate groups to replenish ATP rapidly during energy-demanding activities. This effect is not limited to muscle—it benefits brain function and cognitive clarity as well. A daily dose of 3 to 5 grams of creatine monohydrate is safe and effective for most people, especially when combined with resistance training.

Lastly, **B-vitamins** are foundational. Vitamins B1 (thiamine), B2 (riboflavin), B3 (niacin), B5 (pantothenic acid), and B6 (pyridoxine) are all essential cofactors in the Krebs cycle and electron transport chain. Without adequate levels, even the best mitochondria can't produce energy. Activated forms like benfotiamine (B1), riboflavin-5-phosphate (B2), and P5P (B6) tend to be better absorbed and utilized. A well-rounded B-complex taken in the morning can provide the necessary cofactors for mitochondrial function throughout the day.

Timing matters. Most mitochondrial support supplements are best taken in the first half of the day to avoid overstimulation or sleep disruption. Pairing these interventions with regular movement, quality sleep, and a low-toxicity environment creates a comprehensive ecosystem where mitochondria can thrive.

Restoring mitochondrial health is not about chasing one magic pill. It's about stacking small, deliberate practices that signal to your cells: energy is needed, energy is possible, and the environment is safe to grow. When you nourish your mitochondria, you don't just fight fatigue—you restore your fundamental capacity to be fully alive. You reclaim the ability to focus, move, recover, and dream with clarity and vigor. And in that restoration, you remember that energy isn't something external you need to find. It's something internal you were born with. You simply have to learn how to protect and rebuild it.

Chapter 4: The Lymphatic System – Your Body's Forgotten Detox Highway

4.1 The Lymphatic Network: Your Body's Sanitation System

If your blood is the courier of life, delivering oxygen and nutrients to every cell in your body, then your lymphatic system is its quiet counterpart—the janitor, the recycler, the gatekeeper of immunity, the waste management crew that no one thinks about until something goes wrong. It's ironic, almost absurd, that most people go their entire lives without knowing where their lymph nodes are or what the lymphatic system actually does, considering it processes an estimated three to four liters of fluid every day. Unlike the heart-driven circulatory system, the lymphatic network has no central pump, yet it manages to drain waste from every tissue, clear dead cells, move immune troops into position, and keep our inner landscape clean, balanced, and ready to respond.

The lymphatic system is immense. It's made up of a vast network of vessels, ducts, glands, and nodes that weave through almost every tissue in the body. This system operates largely in the shadows, beneath the skin and around our organs, yet it plays a starring role in maintaining physiological homeostasis. It's composed of thin-walled lymphatic capillaries that collect excess interstitial fluid—the fluid that surrounds your cells and is constantly being exchanged with blood plasma. This fluid, once collected, becomes "lymph," a clear or slightly yellow substance rich in cellular debris, immune cells, proteins, and sometimes pathogens or toxins.

Unlike the circulatory system, which is pressurized by the heart's continuous pumping action, the lymphatic system moves fluid slowly and only in one direction: upward, toward the chest. It relies on a series of **one-way valves** embedded in the vessels, which prevent backflow, and on external forces to move lymph along its path—muscle contractions, deep diaphragmatic breathing, body movements, and even manual stimulation like massage. This means that when you're sedentary for long periods, lymph stagnates. And unlike blood, which recirculates, lymph must be constantly cleared and processed by lymph nodes to avoid a buildup of waste. In other words, motion isn't just good for your muscles—it's necessary for your lymphatic function.

There are between 600 and 700 lymph nodes in the human body, concentrated in key regions like the neck, armpits, groin, abdomen, and chest. These nodes serve as critical filtration points where lymph is cleaned and surveyed. They house vast numbers of immune cells—T cells, B cells, dendritic cells, and macrophages—ready to identify and respond to invaders. Lymph nodes are not just passive checkpoints; they are dynamic command centers. When you get an infection, lymph nodes swell because the immune system is going into overdrive. The nodes are recruiting, replicating, and deploying immune soldiers to contain and eliminate the threat.

This immune aspect is one of the most overlooked but essential functions of the lymphatic system. It doesn't just clear waste—it trains the body to recognize danger. When pathogens enter the body, whether through the skin, respiratory tract, or gut, they often travel through the lymphatic vessels and end up in lymph nodes. There, immune cells sample the invaders, learn their molecular "signatures," and mount appropriate defenses. Without a functioning lymphatic network, the immune system loses its strategic vision. It becomes disorganized, sluggish, or worse—misguided, attacking the body's own tissues.

Nowhere is the importance of lymphatic health more clear than in the **glymphatic system**, a recently discovered mechanism by which the brain clears its own waste. For decades, scientists puzzled over how the brain detoxified itself, since it lacks traditional lymphatic vessels. The answer came only recently: the brain uses a unique structure, dependent on glial cells (hence the term "glymphatic"), that flushes cerebrospinal fluid through brain tissue during deep sleep, clearing away toxic metabolic byproducts like amyloid-beta—the same protein that accumulates in Alzheimer's disease.

This nightly detox cycle is most active during **slow-wave sleep**, the deepest phase of non-REM sleep. During this time, the spaces between brain cells expand, and cerebrospinal fluid is pumped rhythmically through brain tissue, sweeping away cellular waste and carrying it out through meningeal lymphatic vessels recently discovered on the outer layers of the brain. The implication is profound: if you are not getting enough deep sleep, your brain is not clearing waste properly. Over time, this leads not only to cognitive fog and mood instability but potentially contributes to long-term neurodegenerative disease.

The glymphatic system's dependence on quality sleep gives a new meaning to the idea that sleep is restorative. It is not just about rest—it is an essential period of neurological sanitation. Sleep deprivation, poor sleep architecture, irregular sleep cycles, or disrupted circadian rhythms all impair this cleansing process, allowing waste to build up night after night. The result is a chronically inflamed brain, one that becomes more prone to degeneration, poor memory consolidation, anxiety, and emotional dysregulation.

Returning to the broader lymphatic network, the importance of this system in detoxification cannot be overstated. While the liver and kidneys are often heralded as the primary detox organs, they depend on the lymphatic system to bring them cellular debris and foreign matter from the periphery. If lymph flow becomes impaired, toxins accumulate in the interstitial fluid, leading to puffiness, pain, and increased susceptibility to illness. This is particularly relevant in today's world, where exposure to environmental pollutants, microplastics, heavy metals, and endocrine-disrupting chemicals is constant.

Sluggish lymphatic flow manifests in subtle but widespread ways: bloating, unexplained fatigue, recurrent infections, skin breakouts, poor wound healing, and even mood instability. The skin itself is often a reflection of lymphatic health—because when the body cannot eliminate toxins efficiently through internal channels, it tries to purge them through the dermis. Conditions like acne, eczema, and even cellulite may be tied to impaired lymphatic drainage.

Another crucial aspect of lymphatic health is its role in managing fluid balance. The system works in tandem with the circulatory system to maintain the right amount of fluid in tissues. Every day, blood plasma leaks out of capillaries into surrounding tissues, delivering nutrients and oxygen. Roughly 90% of this fluid is reabsorbed directly back into blood vessels. But the remaining 10%—which carries waste products, proteins, and immune cells—must be collected by lymphatic capillaries and returned to circulation via the thoracic duct and right lymphatic duct, which empty into large veins near the collarbones.

If this return system falters, fluid accumulates in tissues, leading to swelling or edema. In extreme cases, like with lymphatic obstruction or after lymph node removal (common in cancer treatments), lymphedema can become a debilitating condition. But even in milder forms, lymphatic stagnation contributes to a feeling of heaviness and malaise that many people describe but cannot quite name. It's not just physical weight—it's toxic burden.

Despite its immense influence on detoxification, immunity, neurological health, and fluid balance, the lymphatic system remains criminally neglected in both conventional medicine and mainstream health education. Rarely is it considered in diagnoses outside of cancer or infection. Yet it is deeply implicated in every chronic inflammatory condition—from autoimmune disorders and metabolic disease to neurological decline and skin problems.

One reason for this neglect is that lymphatic issues don't show up well on standard medical tests. There's no "lymph function panel" in blood work. Lymphatic congestion doesn't reveal itself on an X-ray or a standard MRI. But it shows itself in symptoms that resist explanation and conventional treatment—fluid retention, chronic fatigue, weakened immunity, poor concentration, and a general sense of systemic congestion.

The irony is that supporting the lymphatic system is not complicated. It doesn't require high-tech interventions or expensive pharmaceuticals. It responds beautifully to basic inputs: movement, hydration, breath, manual stimulation, sleep, and alignment with natural rhythms. But before we can take action, we must first reclaim awareness. We must learn to feel this system—this silent current moving under the surface of everything—and understand how deeply our vitality depends on its flow.

The lymphatic system is not secondary. It is not optional. It is not irrelevant. It is the foundation of your body's ability to clean itself, defend itself, and maintain internal harmony. When you restore its function, the body doesn't just detox—it comes alive. Your skin glows. Your mind clears. Your immune system stands ready. Energy flows not only through muscles and nerves but through every channel designed to keep you balanced and protected.

In the chapters to come, we will explore practical strategies to awaken and support this forgotten highway. But for now, understand this: health does not only depend on what you bring in—good food, clean air, nourishing thoughts—but just as much on what you can release. The lymphatic system is how the body lets go. And learning to support its rhythms may be the most powerful act of healing you've never been taught.

4.2 Lymphatic Stagnation: The Root of Chronic Illness

The lymphatic system, for all its essential functions in detoxification, immune regulation, and tissue fluid balance, is uniquely vulnerable to stagnation. Unlike the circulatory system, which has a muscular pump in the form of the heart, the lymphatic system has no central motor. It relies almost entirely on movement—muscular contractions, joint mobility, breathing, and external forces—to push lymph upward against gravity and through its network of vessels and nodes. When this movement is lacking or impaired, the consequences are not subtle. They may begin with fluid retention, general fatigue, or a sense of heaviness in the limbs, but they quickly snowball into more serious terrain: weakened immunity, toxin buildup, inflammation, and eventually chronic disease.

The modern sedentary lifestyle is, without exaggeration, one of the greatest assaults on lymphatic health in human history. Sitting for extended hours—at desks, in cars, on couches—shuts down the mechanical forces that drive lymph. The muscular contractions in the legs and core that help pump lymph through deep vessels in the thighs and abdomen are silenced. Deep diaphragmatic breathing, which helps massage the thoracic duct and stimulate central lymphatic flow, is replaced by shallow chest breathing, often exacerbated by stress or poor posture. As a result, lymph movement slows by as much as 90 percent. This isn't just a theoretical number—it has physiological consequences. When lymph doesn't circulate, cellular waste begins to accumulate in tissues. Proteins and fluids that should have been cleared linger, attracting water and causing swelling. Immune surveillance becomes impaired, increasing susceptibility to infections and delaying recovery. Over time, this stagnation lays the groundwork for inflammatory conditions, hormonal imbalances, and immune dysfunction.

It's not just lack of movement that slows lymph—it's also what we consume. Processed foods, refined sugars, and inflammatory oils alter the internal chemistry of the body in ways that thicken lymph and congest drainage pathways. The lymphatic system, though fluid-based, is highly responsive to changes in viscosity and composition. When systemic inflammation rises—due to poor diet, hidden food intolerances, or even chronic stress—pro-inflammatory cytokines are released into the bloodstream. These signaling molecules, while necessary in acute responses, become pathological when overproduced. They make lymphatic fluid more viscous, harder to pump, and more prone to stasis.

Sugar is a particularly insidious culprit. It feeds dysbiotic bacteria in the gut, promotes insulin resistance, and fuels inflammatory cascades that affect not only blood sugar levels but immune balance and detoxification capacity. High intake of sugar leads to glycation, where sugar molecules bind to proteins, creating sticky residues known as advanced glycation end-products (AGEs). These AGEs can deposit in tissues, contribute to cellular damage, and clog up drainage routes. The lymphatic system must work overtime to clear them—if it can at all. Add in the artificial preservatives, emulsifiers, and chemical additives found in most packaged foods, and the burden becomes overwhelming. The body was not designed to process this level of chemical insult without assistance. And without efficient lymphatic flow, those chemicals linger in the extracellular matrix, altering the function of the very tissues they were meant to nourish.

Food intolerances add another layer to this story. For many people, certain foods—gluten, dairy, soy, or nightshades—trigger low-grade immune responses that create constant inflammation. This kind of reaction often goes unnoticed at first. There may be no overt allergy, no anaphylactic response. But inside, the body is waging a war. Lymph nodes swell. Lymph thickens. Drainage slows. This condition can persist for years, leading to a baseline level of systemic congestion that becomes normalized. People begin to live with symptoms like brain fog, joint stiffness, skin eruptions, or constipation without understanding that the root is lymphatic stagnation caused by their everyday diet.

In some cases, stagnation isn't caused by lifestyle or food but by structural damage. Surgeries, especially those involving lymph node removal or cutting across major lymphatic vessels, can permanently alter lymphatic drainage. Mastectomies, abdominal surgeries, hernia repairs, and even cosmetic procedures can lead to localized lymphedema—persistent swelling and fluid buildup in the affected limb or region. Scar tissue, another common consequence of injury or surgery, compounds the problem. It restricts movement, compresses vessels, and impedes flow. The body may try to compensate by rerouting lymph through nearby vessels, but these collateral pathways are often insufficient.

This kind of regional stagnation is most visible in post-operative patients but can affect anyone who has experienced trauma, repeated strain, or even poor posture over time. A twisted ankle, for instance, can lead to localized lymphatic congestion that lingers long after the acute injury heals. A C-section scar can disrupt lymphatic return from the pelvis. A tight neck and shoulder girdle from hunching over a laptop can compress the deep cervical nodes and affect drainage from the face and head, potentially contributing to sinus congestion, headaches, or jaw tension.

Unfortunately, mainstream medicine often treats these symptoms as isolated or unconnected. Swelling is addressed with diuretics. Skin flare-ups are treated with topical steroids. Infections are managed with antibiotics. Rarely is the lymphatic system even considered, let alone supported. Yet all of these manifestations point to the same underlying issue: fluid is not moving. Waste is not being cleared. The inner highways of the body are jammed, and the resulting traffic causes a cascade of secondary issues.

Understanding lymphatic stagnation as a root cause shifts the way we approach health. It invites us to look at patterns of subtle congestion—physical, emotional, and energetic—and to begin restoring flow. Because when flow is restored, healing can happen. The body is extraordinarily capable of regeneration. But it cannot repair what it cannot clean. And it cannot clean what it cannot move. Stagnation is the soil in which disease takes root. Flow is the antidote.

4.3 Lymphatic Enhancement and Drainage Techniques

Restoring lymphatic flow is not a complicated process, but it does require intention and consistency. The goal is to recreate the natural movements and stimuli that the body evolved with— movements that push lymph against gravity, open up congested pathways, and support the body's innate capacity to cleanse and heal. Unlike pharmacological interventions that suppress symptoms, lymphatic enhancement techniques awaken the body's own systems. They don't force—they encourage. They don't override—they assist.

One of the most effective tools for stimulating lymphatic drainage is manual therapy. **Manual lymphatic drainage (MLD)** is a highly specific massage technique developed to mimic the rhythm and direction of natural lymph flow. It is not deep tissue work. In fact, it relies on light, rhythmic, directional strokes that target superficial lymphatic vessels lying just under the skin. These strokes follow anatomical drainage pathways— always moving toward major lymph node clusters and central ducts. For example, lymph from the legs drains upward toward the inguinal nodes in the groin, while lymph from the arms moves toward the axillary nodes under the armpits. The practitioner's touch must be gentle but deliberate, using the pressure of a nickel or less to avoid collapsing the delicate lymphatic capillaries.

When applied correctly, manual drainage stimulates vessel contraction, opens stagnant capillaries, and encourages the movement of fluid toward filtering nodes. It can be especially helpful for post-surgical recovery, sinus congestion, swelling, and general detox support. But even without a practitioner, self-massage techniques can be learned and practiced at home. Just a few minutes a day of gentle stimulation around the neck, clavicles, and abdomen can awaken stagnant flow and refresh the system.

Movement-based therapies also play a crucial role. The lymphatic system thrives on **rhythmic, vertical motion**, which is why **rebounding**—bouncing gently on a mini-trampoline— has become a popular and effective tool. The gravitational shifts involved in bouncing create alternating compression and decompression in tissues, which mechanically pumps lymph through vessels and into nodes. It's not about intensity or duration—it's about rhythm. Just 10–15 minutes a day of low-impact rebounding can dramatically enhance lymph flow, especially when paired with diaphragmatic breathing.

Another gravity-based approach is **inversion therapy**, where the body is tilted or fully inverted to assist the downward flow of lymph from the legs and lower abdomen toward the thoracic duct. This can be done with inversion tables, yoga poses like shoulder stands, or simply by lying with legs elevated against a wall. These positions not only aid lymphatic return but also reduce pressure in the lower extremities and promote parasympathetic nervous system activation, enhancing rest and repair.

Botanical support has long been used in traditional medicine systems to assist lymphatic health. Certain herbs possess affinities for lymphatic tissues, encouraging drainage, reducing swelling, and supporting immune function. **Cleavers** is one such herb, traditionally used to "thin" lymphatic fluid and promote movement. It is often prepared as a tincture or infusion and taken daily during periods of congestion. **Red root**, also known as *Ceanothus americanus*, is a powerful astringent and lymphatic decongestant, particularly useful in cases of swollen lymph nodes or tonsils. **Calendula**, gentle but effective, supports lymphatic movement and soothes inflammation.

These herbs can be used alone or in combination, often as part of a cleansing protocol. But they are most effective when supported by external practices like **castor oil packs**—a traditional remedy where castor oil is applied to the skin over lymph-rich areas like the abdomen or chest, then covered with a cloth and warmed. The oil penetrates deeply into tissues, reducing inflammation, dissolving adhesions, and promoting lymphatic flow. It is particularly useful for hormonal imbalances, digestive stagnation, and liver support.

Dry brushing is another simple but powerful technique that stimulates lymph just beneath the skin. Using a natural-bristle brush, the skin is gently stroked in the direction of lymphatic drainage—always toward the heart. When done before a shower, it not only boosts circulation and exfoliates the skin but also wakes up the entire lymphatic system, preparing the body for detox.

None of these techniques require advanced knowledge or expensive tools. What they require is presence—a willingness to engage with your body's rhythms and to support its efforts to stay clean, clear, and vital. The lymphatic system, more than most, responds to this kind of care. It does not need to be fixed—it needs to be allowed to move.

In a world that constantly bombards us with toxins, stress, and artificial inputs, lymphatic care is one of the simplest and most profound ways to reclaim autonomy over our health. It reminds us that healing doesn't always come from outside. Often, it comes from restoring the ancient flows that already exist within us. It comes from remembering that detox is not a dramatic event—it is a daily rhythm. And the more we align with that rhythm, the more vitality, clarity, and resilience we reclaim. The body was designed to cleanse itself. Our job is to get out of its way—and to support it in the places it has been blocked. That is the promise of lymphatic renewal. And it begins with movement.

Chapter 5: Fascia – The Body's Internet Network That Modern Medicine Ignores

5.1 The Fascial Web: Your Body's Structural Internet

If you've ever stretched a tight hamstring, rubbed a sore neck, or received a deep tissue massage, you were interacting with one of the most intelligent and overlooked systems in your entire body—your fascia. For decades, fascia was dismissed by modern medicine as inert packing material, little more than the white connective tissue technicians sliced through to access "more important" structures during dissection. But today, the picture is radically shifting. What was once viewed as anatomical fluff is now recognized as one of the most pervasive and dynamic communication systems in the body—a structural, biochemical, and sensory matrix that connects every organ, muscle, and cell. And it's far more than structural. It senses, it remembers, it responds, and it influences everything from posture and pain to emotion and immunity.

The fascia is not just one structure, but rather a vast, uninterrupted network of connective tissue that wraps around and weaves through muscles, bones, nerves, and organs. It exists in layers and strata, with deep fascia enclosing muscle groups, visceral fascia supporting internal organs, and superficial fascia linking the skin to underlying structures. From head to toe, it's one continuous system. This interconnectedness means that tension, injury, or restriction in one area can ripple across distant regions—creating effects far from the original point of dysfunction. In this sense, fascia does not simply "surround" the body's structures. It *is* the body's structure.

One of the most elegant ways to understand fascia is through the concept of **tensegrity**—a term that combines "tensional integrity." In tensegrity systems, integrity is maintained not through rigid stacking of parts, like bricks in a wall, but through a balance of continuous tension and local compression. Imagine a suspension bridge or a geodesic dome. The structure is held together by the balanced tension in cables, not by the mass of heavy materials. The human body, it turns out, follows the same principle. Your bones do not bear your weight like pillars; they float in a tensional matrix of fascial tissue. It's the fascia that creates the scaffolding that suspends and aligns them.

This architectural model allows the body to be simultaneously strong and flexible, responsive and resilient. It also explains how force is distributed. When you strike your foot against the ground, the impact doesn't just stop at your ankle. It travels upward, through your leg, pelvis, spine, and shoulders. Fascia allows these forces to be absorbed, redirected, and dissipated across the entire body. This is why a knee problem might originate in the hip. Or why tension in the jaw can affect the lower back. Fascia doesn't operate in isolation—it's a web. And a web doesn't tear in one strand without affecting the whole.

But fascia isn't only mechanical. It's biological. The **extracellular matrix** (ECM) that makes up the fascial system is a vibrant, hydrated, gel-like substance composed of collagen fibers, elastin filaments, and a ground substance made primarily of water, glycoproteins, and proteoglycans. This matrix provides more than structural support—it's a medium for intercellular communication. Through it, nutrients are delivered, waste products are removed, and chemical signals are transmitted. The health of your ECM influences how well your cells breathe, metabolize, divide, and repair. A dehydrated or stagnant matrix impedes all these processes, leading to chronic inflammation, slow healing, and tissue degeneration.

Collagen, the most abundant protein in the body, is a primary component of fascia. It's what gives fascia its strength and integrity. But collagen is not static—it's constantly being broken down and rebuilt in response to demand and stimulus. This remodeling depends on mechanical input, hydration, and adequate nutrient support, particularly vitamin C, amino acids, and certain trace minerals. Elastin, though present in smaller amounts, adds elasticity, allowing fascial tissue to stretch and recoil. Together, these components form a living suit of armor—capable of incredible adaptability when cared for, and equally capable of stiffening, shrinking, and entangling when neglected.

The ground substance—the space between the fibers—is where much of the fascia's intelligence resides. In a healthy state, it has a viscous, gel-like consistency that permits smooth gliding between tissue layers. But under stress, inflammation, or dehydration, it thickens and becomes adhesive. Layers that should slide instead stick. Movement becomes restricted. This leads to compensatory patterns, postural imbalances, and pain. More than that, it interferes with the subtle, often electrical forms of communication that take place within this matrix—communications that influence not only muscles but immune cells, nerve endings, and even gene expression.

That brings us to one of fascia's most remarkable and least understood functions: **mechanotransduction**. This is the process by which mechanical forces—pressure, stretch, compression—are converted into biochemical signals that influence cellular behavior. In other words, when you move, stretch, compress, or even gently manipulate your fascia, you're not just affecting muscles—you're sending messages to your cells. Messages that can turn genes on or off, influence protein synthesis, and even dictate how cells differentiate and repair.

This mechanotransductive process is possible because of specialized structures in the fascia called **integrins**—

transmembrane proteins that connect the outside environment (the ECM) to the cytoskeleton of the cell. When the ECM is deformed—by movement, massage, posture, or trauma—these integrins translate the mechanical signal into a cascade of intracellular events. This influences everything from inflammation and immune response to stem cell activity and tissue regeneration. It also helps explain why physical therapy, bodywork, yoga, and stretching can have systemic effects that go far beyond the obvious range of motion improvements. You're not just mobilizing joints—you're reprogramming your biology.

Fascial mechanotransduction is not hypothetical. It has been observed in lab settings where cells cultured on stiffer matrices behaved differently from those on more flexible ones, despite identical genetics. In the human body, it suggests that chronic fascial tension—whether due to injury, poor posture, emotional trauma, or repetitive stress—can shape not only how we move but how we heal, how we age, and how we experience pain. Chronic tension alters the biochemistry of the fascia, increases fibroblast activity, and may even lead to fibrotic changes that entrap nerves and restrict circulation.

The sensory capacity of fascia further elevates its importance. Recent studies have revealed that fascia contains far more sensory nerve endings than previously thought—some estimates place it at ten times the sensory density of muscle tissue. These nerve endings include proprioceptors (which detect position and movement), nociceptors (which sense pain), and interoceptors (which monitor internal states). This makes fascia not just a structural or mechanical tissue but a sensory organ. It's the tissue through which we *feel* ourselves—our shape, our space, our presence in the world.

This sensory function also plays a role in emotional regulation. Fascia is deeply connected to the autonomic nervous system, particularly the vagus nerve. Chronic fascial tension often mirrors chronic emotional tension—anxieties stored in the gut, grief in the chest, stress in the neck and shoulders. When fascia tightens, it creates mechanical stress that keeps the nervous system in a state of hypervigilance. Releasing that tension—whether through breathwork, stretching, myofascial release, or somatic therapies—can downregulate the stress response and restore a sense of safety and ease in the body.

It's not a coincidence that many people experience emotional releases during bodywork sessions. The fascia, saturated with sensory nerves and intimately tied to our sense of self, holds memory—not just of trauma or injury, but of movement, stillness, and felt experience. It is a tissue of history as much as structure. To work with fascia is to communicate with the body's deepest layers—its architecture, yes, but also its memory, its identity, its aliveness.

Despite all this, fascia remains one of the most underrepresented subjects in medical education. Most physicians graduate with little more than a passing mention of connective tissue. And yet, an increasing number of clinicians, therapists, and researchers are realizing that this network holds answers not found in blood tests or scans. Fascia does not show up clearly on standard imaging, which is part of why it has been so long ignored. But dysfunction in this tissue explains so many "mystery" symptoms—pain that doesn't follow nerve patterns, tightness that resists stretching, fatigue that is not caused by weakness but by restriction and inefficiency in movement.

The modern body, increasingly sedentary, stressed, and tech-bound, is particularly prone to fascial dysfunction. Long hours in static postures, repetitive movements, and chronic underhydration create an internal terrain where the fascia thickens, dehydrates, and loses its elasticity. Layers that should glide instead stick. Signals that should flow instead stagnate. Over time, the body loses its fluidity, its resilience, its ability to adapt. It becomes rigid—not just physically, but emotionally, mentally, energetically.

But this can be reversed. Fascia is responsive. It adapts quickly to new stimuli. With consistent movement, hydration, and intentional care, it regains its fluidity. The glide returns. The stiffness dissolves. The nervous system calms. Pain subsides. And perhaps most importantly, the person reconnects with their own body—not as a machine to be fixed, but as a network to be felt, listened to, and lived in.

Fascia is not just connective tissue. It is the body's communication network. Its internal internet. Its shape, its memory, and its wisdom. To ignore it is to miss the thread that links structure to sensation, anatomy to experience, biology to consciousness. But to understand it—even a little—is to begin healing in a way that is not only structural, but holistic. In the next section, we'll explore how to care for this system—how to hydrate it, stretch it, release it, and restore its intelligence. But for now, it's enough to know this: the fascial web is not a passive background. It is the living matrix of your being. And it's listening.

5.2 Fascial Dysfunction: The Hidden Source of Chronic Pain

Chronic pain is one of the most elusive conditions in all of medicine. It lingers where injuries have long healed, appears without a clear cause, and resists conventional treatments like anti-inflammatories or muscle relaxants. In many cases, imaging reveals nothing—no structural tear, no nerve impingement, no arthritic degeneration. Yet the pain is real, persistent, and often debilitating. This medical mystery has led countless people through endless appointments, misdiagnoses, and dismissive labels like "psychosomatic" or "fibromyalgia" with no explanation of what's actually wrong. One of the reasons chronic pain is so poorly understood is that the system responsible for much of it—**the fascial system**—has remained largely invisible to traditional diagnostic tools and medical thinking. But fascia, when injured, dehydrated, or emotionally burdened, can be a profound and silent generator of pain.

Fascial dysfunction often begins subtly. It doesn't require a major accident or trauma. Repetitive movements, poor posture, surgical scars, or even emotional stress are enough to trigger small disruptions in the fascial web. When fascia is stressed or injured, it undergoes a biological change. Instead of remaining supple and fluid, its fibers begin to thicken and bond together, forming **adhesions**—cross-links between layers that are normally supposed to glide smoothly. These adhesions act like knots in a spider web. They create restrictions that limit movement, redistribute force improperly, and increase mechanical stress on nearby joints and tissues.

Over time, these restrictions trigger a domino effect. The body, always trying to function efficiently, compensates. It shifts posture. It recruits other muscles. It alters movement patterns. But compensation always comes at a cost. Areas that should be relaxed become overactive. Areas that should bear weight lose strength. Muscles become imbalanced, joints become misaligned, and the fascia becomes more bound. What started as a small adhesion in the shoulder, for example, can eventually contribute to neck stiffness, elbow tendonitis, or even low back pain. This is why treating the site of pain often yields limited results—the problem rarely resides where it hurts.

This leads to one of the most puzzling aspects of fascial dysfunction: the **fascial pain paradox**. Unlike the nervous system, which follows relatively predictable pathways, fascia transmits force and tension across vast, non-linear distances. These pathways, known as myofascial meridians or anatomy trains, crisscross the body in spirals and chains that defy traditional textbook anatomy. A restriction in the plantar fascia of the foot can influence the hamstrings, sacrum, and even the cervical spine. A scar from an abdominal surgery can create tension that pulls on the diaphragm, affecting breathing, posture, and shoulder mobility. This web of tension explains why people often experience pain far from the actual source of dysfunction. The symptom may appear in one location, but the cause is hidden upstream in the fascial matrix.

Pain from fascial restriction is also different in quality. It's often described as dull, aching, burning, or diffuse rather than sharp or localized. It can shift, change intensity, or appear in strange patterns. Sometimes it intensifies with inactivity—particularly after long periods of sitting or sleeping—and eases with movement, only to return later. In some cases, it mimics nerve pain without any nerve impingement. In others, it overlaps with what is traditionally diagnosed as tendinitis or bursitis, yet fails to respond to anti-inflammatory treatments. These inconsistencies frustrate both patients and practitioners. But when viewed through the lens of fascia, they make sense. A tensegrity-based body cannot be understood by looking at one isolated part. The whole system must be considered.

Fascial dysfunction is not just mechanical—it's biochemical and emotional, too. Fascia responds to inflammation by changing its molecular makeup. The ground substance between the collagen fibers becomes thicker, more viscous. This loss of fluidity means that layers stick together rather than glide. Nutrient exchange slows. Cellular waste builds up. The environment becomes more hostile to proper function. Chronic dehydration exacerbates this process. Fascia is highly hydrophilic—it draws and holds water to maintain its gel-like state. But when the body is under-hydrated, either from poor water intake or imbalanced electrolytes, fascia loses its pliability. It becomes more fibrous, more rigid. And when fascia stiffens, movement becomes restricted, circulation suffers, and pain thresholds drop.

A dehydrated fascial matrix isn't just a dry structure—it's a compromised communication medium. The extracellular matrix is where many of the body's messages are sent and received. Hormones, immune signals, and neurotransmitters all travel through this fluid highway. When it thickens or loses integrity, these signals are distorted. This can contribute to systemic issues like fatigue, brain fog, and heightened sensitivity to pain. It may even explain aspects of "central sensitization," where the nervous system overreacts to stimuli that wouldn't normally cause pain.

Moreover, fascia appears to play a role in the storage of emotional memory. While this might sound like a poetic metaphor, growing evidence and countless clinical observations suggest otherwise. Emotional trauma often manifests as muscular bracing—subtle, unconscious tension patterns that linger long after the threat is gone. These patterns become embedded in the fascial tissue, like software running silently in the background. Over time, the body forgets how to fully relax. The tissue adapts to tension as its new normal. This can create chronic pain loops that are no longer tied to physical injury, but to unprocessed emotional stress. Releasing these patterns sometimes brings not

71

just physical relief, but emotional release—a sense of catharsis that transcends biomechanics.

Traditional medicine, with its emphasis on linear cause-and-effect, struggles to map this kind of systemic, interconnected dysfunction. Pain is usually treated symptomatically—with medications, injections, or surgeries aimed at the site of discomfort. But when the root lies in bound fascia, these approaches are often ineffective or temporary at best. Understanding the fascia means stepping into a more holistic paradigm—one where the body is not a collection of parts, but a unified network of communication, structure, and sensation. And when that network is stuck, pain is often the body's way of asking to be heard.

5.3 Fascial Release and Restoration Strategies

The good news is that fascia is remarkably adaptable. It's not static scar tissue—it's living, responsive tissue that can change with the right input. Once restrictions are identified and addressed, the fascial web can soften, rehydrate, and realign. Pain fades. Movement becomes fluid. The nervous system calms. But unlike muscles, which respond well to stretching and strengthening, fascia requires a different approach—one based on sustained pressure, varied movement, and consistent hydration.

One of the most effective self-care tools for releasing fascial tension is **myofascial release**. Unlike conventional massage, which often targets muscle bellies with deep, rhythmic strokes, myofascial release focuses on applying gentle but sustained pressure into restricted fascial areas. The goal is not to "push through" tension but to sink into it slowly, allowing the tissue to soften from within. Foam rollers, lacrosse balls, and specialized tools like massage canes or therapy sticks can be used effectively across different parts of the body. But technique matters. Rolling too quickly or applying too much pressure can actually cause the tissue to brace, reinforcing tension rather than releasing it.

For optimal effect, pressure should be applied slowly and held for at least 90 to 120 seconds in one spot, allowing the fascia to yield. Movements should be slow and deliberate, often combining compression with gentle shearing or cross-fiber strokes. Different regions require different tools—larger rollers for the thighs and back, smaller balls for the feet, shoulders, and glutes. And because fascia is a system, it's important to treat not just the site of pain but the surrounding lines of tension. Releasing the calves, for instance, can have a surprising impact on lower back mobility. Working on the chest and diaphragm can improve shoulder function and breathing. Over time, consistent self-treatment creates space in the system. Layers begin to glide again. Posture shifts. Pain dissolves.

Movement is another powerful way to restore fascial health, especially movement that goes beyond linear repetitions. The fascial system thrives on **diverse, multidirectional motion**. Unlike muscles, which contract in straight lines, fascia stretches and responds in spirals, waves, and diagonals. This is why traditional exercise, when too rigid or repetitive, can sometimes contribute to fascial imbalance. Walking on flat surfaces, lifting in predictable patterns, or sitting in symmetrical postures can create blind spots in the body's movement vocabulary. Fascial training reintroduces variety. Practices like animal flow, functional mobility, tai chi, and certain forms of dance emphasize spiraling, bouncing, and fluid transitions that awaken dormant pathways and hydrate connective tissue.

Fascial elasticity is also enhanced through **oscillatory movements**—gentle bouncing, rocking, or pulsing that mimic the natural rhythms of breath and heartbeat. These micro-movements stimulate mechanoreceptors in the fascia, reawaken proprioception, and improve tissue hydration. Think of it like squeezing a sponge and then letting it reabsorb water. This pulsing encourages circulation, lymphatic drainage, and neuromuscular reconnection. It's especially effective for people with stiffness, chronic fatigue, or pain hypersensitivity, as it reintroduces movement without overwhelm.

None of this works, however, without proper **hydration**. Fascia is made of water-bound gel. Without sufficient fluid, its gliding properties disappear. But hydration is not just about drinking more water—it's about absorbing and retaining it at the cellular level. This requires adequate electrolytes, particularly sodium, potassium, and magnesium, which help water move across cell membranes. Timing also matters. Sipping water throughout the day is more effective than guzzling large amounts at once. Movement, again, plays a role—hydration gets "locked in" when fascia is compressed and released, much like a sponge.

Nutritionally, several compounds have been shown to support fascia. **Hyaluronic acid**, found in connective tissue and synovial fluid, promotes hydration and viscosity. It can be taken as a supplement or found in bone broth, organ meats, and fermented foods. **Vitamin C** is essential for collagen synthesis and repair—without it, the fascial web weakens and becomes more prone to injury. **Silica**, a trace mineral found in plants like horsetail, bamboo, and leafy greens, helps maintain the integrity of connective tissue and supports collagen cross-linking. These nutrients, when combined with movement and hydration, create the internal conditions for fascial regeneration.

Restoring fascial health is not a single protocol or one-size-fits-all fix. It's a practice—a conversation between body and mind, movement and stillness, tension and release. It's about listening for what feels bound and finding ways to bring it back into motion. And in doing so, pain becomes less of a mystery and more of a messenger—pointing not to something broken, but to something that wants to move again. The body, when given the chance, always moves toward wholeness. The fascia just needs to be shown the way.

Chapter 6: The pH Balance Myth and Acid-Base Reality

6.1 Understanding True Acid-Base Physiology

The idea that we can manipulate our health by simply "alkalizing" the body has become one of the most enduring wellness trends of the last few decades. The market is flooded with alkaline waters, pH testing strips, green powders, and strict food lists dividing the "acidic villains" from the "alkaline heroes." Influencers preach the dangers of acidic diets, linking them to everything from cancer to fatigue. It's a compelling story. It feels scientific. But it's also misleading. In reality, your body maintains a tightly regulated blood pH within the range of 7.35 to 7.45—and if that shifts by even half a point in either direction, you're in a medical emergency. You don't walk it off. You don't "alkalize" your way back. You go to an ICU.

The truth is both more complex and far more fascinating than the wellness industry would have you believe. Acid-base balance in the human body is not a single number. It's a highly dynamic, compartmentalized, and electrically regulated process that operates across tissues, fluids, and time scales. Different parts of the body require vastly different pH levels to function. Your stomach, for instance, must be extremely acidic—around 1.5 to 2.5 pH—to digest protein and kill pathogens. Meanwhile, your pancreatic secretions are deliberately alkaline, often reaching 8.5, to neutralize that acid before it enters the small intestine. Your skin is slightly acidic, while your blood is tightly buffered. The variation is intentional and essential.

Maintaining this multi-compartmental pH landscape requires intricate biological machinery—much of which operates far beyond dietary choices. That doesn't mean food doesn't matter. But it means the real story of pH involves buffering systems, cellular voltage, respiratory patterns, kidney function, and even electromagnetic fields—not just whether you drank lemon water this morning.

At the heart of the body's pH regulation are **buffering systems**—physiological shock absorbers that prevent sudden changes in pH. The most well-known of these is the **bicarbonate buffer system**, which operates primarily in the blood. When acid accumulates, bicarbonate (HCO_3^-) binds with hydrogen ions (H^+) to form carbonic acid (H_2CO_3), which then quickly converts to water and carbon dioxide. The CO_2 is exhaled by the lungs, completing the neutralization process. This system works within seconds, offering immediate protection against pH swings caused by metabolism, exertion, or emotional stress.

But the bicarbonate buffer is only one piece of the puzzle. The **phosphate buffer system** plays a key role in intracellular environments and in the kidneys. It helps regulate the pH of urine and supports the excretion of hydrogen ions without disturbing other mineral balances. Then there's the **protein buffer system**, which involves amino acids in blood plasma and within cells. Proteins, particularly hemoglobin in red blood cells, have both acidic and basic groups, making them ideal candidates for buffering. Hemoglobin, in fact, not only carries oxygen but also mops up hydrogen ions generated by the release of carbon dioxide from tissues.

These systems don't operate in isolation. They are part of a finely tuned orchestra, keeping each region of the body within its necessary pH zone. And within cells, this regulation gets even more granular. Cellular pH can differ significantly from blood pH, and it fluctuates with energy production, signaling cascades, and enzyme activity. For example, many enzymes involved in ATP synthesis operate best within a narrow pH range inside the mitochondria. Slight shifts can speed up or slow down metabolic processes dramatically. That's not just about "acid" or "alkaline"—it's about bioelectric coherence.

This is where **cellular voltage** comes into play. Every cell in your body functions like a miniature battery. The voltage across the cell membrane—often measured as a membrane potential in millivolts—is generated primarily by the **sodium-potassium pump**, a protein that uses ATP to move sodium ions out of the cell and potassium ions in. This creates an electrochemical gradient, essential for nutrient transport, electrical signaling, and maintaining intracellular pH.

The movement of hydrogen ions, in particular, is tightly regulated, because these ions determine acidity. Specialized proton pumps and ion exchangers control their flow. When the sodium-potassium pump operates efficiently, the cell maintains a stable charge and can manage its internal environment. But when cellular voltage drops—due to inflammation, nutrient deficiency, toxin exposure, or mitochondrial dysfunction—the cell struggles to maintain homeostasis. Intracellular pH may drift. Enzymes misfire. Detoxification slows. Communication breaks down.

So while diet can influence the body's acid load to a degree—particularly in the short term—it is your cells' **electrical integrity** that determines how well that acid load is processed and eliminated. This is why two people can eat the same "acidic" meal and respond very differently. One has strong cellular voltage, robust mineral reserves, and efficient buffering systems. The other does not. The difference isn't pH alone—it's metabolic adaptability.

Another major pillar of pH regulation is the **respiratory system**. Every time you exhale, you release carbon dioxide—a weak acid when dissolved in blood. The respiratory center in the brainstem constantly monitors blood pH and adjusts breathing rate accordingly. If you're too acidic, your breathing speeds up to blow off more CO_2. If you're too alkaline, it slows down to retain it. This system responds in minutes and can compensate for metabolic imbalances rapidly.

Breathing patterns, therefore, have a direct influence on pH. Hyperventilation, whether due to stress or shallow overbreathing, can drive off too much CO_2 and cause **respiratory alkalosis**. This leads to symptoms like dizziness, tingling, and anxiety—not because you lack oxygen, but because your blood has become too alkaline. On the other hand, hypoventilation—such as what occurs during sleep apnea, chronic lung disease, or shallow, inhibited breathing—can cause **respiratory acidosis**, where CO_2 builds up and the blood becomes more acidic. Both states disrupt cellular function and, over time, contribute to systemic stress.

The **kidneys** provide a slower but more sustained method of pH regulation. They filter blood plasma, excreting hydrogen ions and reabsorbing bicarbonate to maintain balance. This process takes hours to days, but it's essential for long-term acid-base stability. Unlike the lungs, which handle volatile acids like CO_2, the kidneys deal with fixed acids—those generated by protein metabolism, such as sulfuric and phosphoric acid. They also manage the reabsorption and excretion of minerals like calcium, magnesium, and potassium, which are often mobilized during acid-base compensation.

This is where prolonged dietary patterns begin to matter more. Diets high in animal protein, processed grains, and low in fruits and vegetables can increase the acid load on the kidneys, particularly if mineral intake is insufficient. The body must buffer this acid somehow, and if bicarbonate stores are low, it may draw alkaline minerals from the bones—especially calcium. This is one of the mechanisms by which poor dietary patterns may subtly contribute to **osteopenia or osteoporosis** over time. But this isn't a matter of eating only "alkaline" foods. It's about **balance, mineral density, and supporting the organs that regulate pH**, not overriding them with dietary extremes.

The common claim that lemon juice or baking soda can "alkalize" the blood is a fundamental misunderstanding of physiology. The stomach must remain acidic to digest food and protect against pathogens. Introducing alkalizing agents indiscriminately can disrupt digestive chemistry, impair protein breakdown, and alter microbial balance. Moreover, the idea of testing urine or saliva pH as a proxy for blood pH is unreliable at best. These outputs reflect **what the body is excreting**, not what the blood is doing. They are compensations, not conditions.

The body's acid-base status is not about one number. It's a **dance of systems**, orchestrated through buffers, respiration, renal excretion, cellular membrane integrity, and mineral balance. The extracellular and intracellular spaces must remain in careful communication. Electrical gradients must be preserved. Enzymes must operate within their optimal pH ranges. The immune system, too, responds to pH, as does microbial ecology. Your gut, your brain, your skin, your mitochondria—all live within narrow windows of acidity and alkalinity, each one calibrated for its function.

What throws this balance off is not a single hamburger or a glass of wine. It's the cumulative stress of poor breathing, sleep deprivation, mineral depletion, chronic inflammation, environmental toxins, sedentary behavior, and poor cellular energy. These stressors weaken the body's buffering systems and reduce its capacity to adapt. That's when acid load becomes a problem—not because you've violated a dietary pH rule, but because the underlying systems can no longer compensate.

So the real conversation is not about chasing "alkalinity" as a destination. It's about **supporting the architecture that maintains pH across systems**, in real time, every day. That means breathing deeply, sleeping fully, eating mineral-rich whole foods, staying hydrated, and maintaining strong mitochondrial output. It means understanding that acid and base are not enemies but dual forces in a living system. Acid helps you digest, detoxify, fight infection, and generate energy. Alkaline conditions promote repair, stability, and resilience. Both are necessary. Both must be in flow.

The pH myth persists because it promises a simple solution to complex problems. But biology rarely deals in simplicity. It deals in nuance, feedback loops, and systems intelligence. And nowhere is this more evident than in acid-base regulation. To truly understand health at this level is to understand the intelligence of the body itself—not as a battlefield of acid versus alkaline, but as a symphony of dynamic equilibrium. The blood remains in balance not because we control it, but because the body *already knows* how to keep us alive—so long as we stop interfering and start supporting.

6.2 The Mineral-pH Connection That Changes Everything

While mainstream discussions of acid-base balance tend to fixate on the blood's tight pH regulation, the truth is that pH dynamics occur at a much deeper level—within tissues, cells, and organ systems, driven primarily by mineral-mediated electrical gradients. The real levers of acid-base physiology lie in how the body manages mineral ions like sodium, potassium, calcium, and magnesium. These ions are not just nutrients. They are charged particles that generate and maintain cellular voltage, dictate membrane potential, regulate enzymatic activity, and create the bioelectrical environment in which life unfolds.

Each cell in your body is a miniature battery. It operates across a membrane with a differential charge—inside is typically more negative than outside. This charge differential is not incidental; it is actively created by transporters like the sodium-potassium ATPase pump, which moves three sodium ions out of the cell and two potassium ions in, using energy in the form of ATP. This gradient does more than power nerve impulses. It governs the movement of water, the uptake of nutrients, the elimination of waste, and the ability of the mitochondria to generate energy. Disruption of this mineral balance, especially through chronic stress, processed food consumption, and environmental toxicants, can collapse this cellular voltage and drive dysfunction.

These mineral gradients are also central to regulating intracellular pH. While blood pH is maintained between 7.35 and 7.45 with almost militaristic precision, the pH inside cells varies depending on metabolic activity, organ function, and stress levels. For example, cells with higher energy demand—such as neurons or muscle fibers—must constantly buffer acidic byproducts of metabolism like hydrogen ions. Here, minerals serve as both buffers and regulators. Magnesium acts as a cofactor for hundreds of enzymes that manage acid-base chemistry. Potassium regulates cellular hydration and pH. Calcium plays a complex role in signaling pathways, some of which are directly pH sensitive. When these minerals are depleted, buffering falters, voltage drops, and the internal environment of the cell becomes increasingly acidic, setting the stage for inflammation and degeneration.

The pH requirements of different tissues vary dramatically, but all depend on these finely tuned mineral mechanisms. The skin, for example, maintains an acidic mantle (around pH 4.5 to 5.5) that protects against microbial invasion. The stomach generates hydrochloric acid at a pH below 2 to digest proteins and sterilize food, while the small intestine requires a more alkaline environment (pH 7.5 to 8.5) for pancreatic enzymes to function correctly. Saliva, tears, and urine all operate within specific pH windows that shift in response to hydration status, food intake, and metabolic load. In each of these cases, it is minerals—not alkaline water or green powders—that determine the buffer capacity and setpoint of these fluids.

Modern lifestyles, unfortunately, are engineered to destroy this mineral foundation. The soil our food grows in is depleted. Agricultural practices have stripped essential elements from crops, especially magnesium and trace minerals like zinc and selenium. Food processing removes what little remains. Water purification often eliminates beneficial minerals along with pathogens, leaving us with demineralized, dead water. On top of this, common medications such as diuretics, antacids, and antibiotics alter mineral absorption or cause urinary loss of key ions. Even chronic stress, by elevating cortisol, shifts mineral distribution and increases the urinary excretion of magnesium and potassium, weakening the buffering systems at the cellular level.

Mineral loss is not theoretical—it is epidemic. Studies show that up to 80% of the population is deficient in magnesium, and widespread potassium deficiency is a known contributor to hypertension and insulin resistance. These imbalances do not always show up on standard blood tests because the body will sacrifice tissue reserves to maintain blood levels. But this comes at a cost: reduced intracellular concentrations, impaired mitochondrial function, lower cellular voltage, and the slow slide toward acidity, inflammation, and degeneration.

To restore acid-base balance at the cellular level, we must think in terms of electrical potential, not just pH numbers. It is the charge separation across membranes, driven by mineral gradients, that maintains the integrity of life. When this voltage drops, due to mineral depletion or chronic metabolic stress, cellular communication falters, detoxification slows, and repair mechanisms stall. Without adequate minerals, the body cannot perform the very processes that keep pH stable in the first place.

The mineral-pH connection is not just a chemical affair—it is a bioelectrical necessity. Your health depends on maintaining the voltage potential across trillions of membranes simultaneously. And the key to sustaining that charge lies not in dietary alkalinity myths, but in providing your body with the mineral raw materials it evolved to use over millennia. Electrolytes are not trendy supplements—they are the primary language of life.

6.3 Optimizing Cellular pH and Mineral Status

Once we understand that pH is a function of cellular charge, and that charge is generated by mineral ion movement, the logical next step is restoring the conditions that allow the body to recalibrate this dynamic. Optimizing pH is not about aiming for an "alkaline body" or chasing an abstract number on a test strip. It's about restoring voltage, flow, and mineral availability in real time, every day. The path toward that optimization starts with replacing what modern life has depleted and aligning lifestyle practices with the rhythms and needs of human physiology.

A foundational step is targeted mineral replenishment. It's not enough to take a generic multivitamin or eat the occasional leafy green. The form, ratio, timing, and context of mineral intake determine how well they restore cellular function. Magnesium, for instance, is better absorbed as glycinate, malate, or threonate rather than oxide. Potassium should be balanced with sodium—not simply restricted—because both work in concert across membranes. Calcium must be bioavailable but not excessive, and ideally offset by sufficient magnesium. Trace minerals like boron, chromium, and selenium, though required in minuscule amounts, regulate critical enzyme systems and hormone pathways. Even sulfur, often overlooked, is essential for detoxification and tissue repair.

To truly support pH at the cellular level, minerals must be delivered in a bioavailable matrix and consumed with attention to the body's natural rhythms. Taking minerals with food enhances absorption. Timing magnesium before sleep can improve parasympathetic tone. Electrolyte solutions in the morning can support hydration and adrenal function after a night of repair. But more than supplementation, this strategy demands a consistent, whole-food diet that is rich in unprocessed plants, wild protein, mineral-rich salts, and clean water.

Breathing practices are another overlooked but powerful lever for pH regulation. Carbon dioxide is not just a waste gas—it is a critical acidifying agent that influences blood pH within seconds. When we overbreathe—taking short, shallow, or rapid breaths—we blow off too much CO_2, causing respiratory alkalosis and reducing oxygen delivery to tissues. Methods like the Buteyko method, which trains the body to tolerate higher CO_2 levels, help normalize pH and enhance oxygen utilization. Coherent breathing techniques, typically around five to six breaths per minute, activate the parasympathetic nervous system, stabilize heart rate variability, and modulate the acid-base setpoint in real time. These techniques, practiced daily, can have profound effects on energy, clarity, and immune resilience.

Water quality may be the most underestimated influence on pH and mineral balance. The body is composed of approximately 70% water, but not all water is equal. Municipal water is often stripped of beneficial minerals and can carry trace contaminants that alter physiology. Bottled water, especially in plastic, may contain endocrine disruptors and lack structure. Optimal hydration comes not just from quantity, but from quality—from water that contains the full spectrum of electrolytes in their natural ratios and retains the molecular coherence found in natural springs. Water that has passed through rocks, picked up trace minerals, and vortexed in nature carries electromagnetic properties that affect how it hydrates cells and supports voltage. Investing in high-quality water filtration systems that preserve mineral content or re-mineralizing your water with trace mineral drops is more than a wellness upgrade—it is a foundation for cellular function.

Together, these practices—targeted mineral restoration, breathwork, and water optimization—form a triad of support for healthy pH regulation. But the deeper truth is that they are not new. They are reminders of how the human organism evolved. We are electrical beings powered by mineral currents, regulated by breath, and hydrated by structured water. When we live in alignment with these principles, pH balances itself—not just in the blood, but in the tissues where disease begins and healing takes root.

Healing pH dysregulation, then, is not a matter of suppression or micromanagement. It is a rebalancing act. A return to an environment—internally and externally—that supports charge, flow, and signal clarity. Acidic tissues, inflamed organs, and chronically fatigued systems are not broken. They are offline. Starved of the voltage they need to function. And the tools to bring them back online are surprisingly simple: minerals, breath, water, rhythm.

The myth of acid versus alkaline has distracted us for too long. The real story is electrical. It is mineral-driven. It is rhythmic, responsive, and utterly within your control. The solution was never in the supplement aisle—it was in your mitochondria, your membranes, and your breath. And now that you know how they work, you also know how to bring them back to life.

Chapter 7: Hormonal Hierarchies - The Endocrine Symphony Playing Out of Tune

7.1 The Hormonal Command Structure Nobody Explains

Your body is a miraculous symphony of chemical messengers, each hormone playing its part in a tightly choreographed rhythm that maintains health, vitality, and equilibrium. Yet unlike a free-flowing jazz band, your endocrine system isn't democratic. There's a strict chain of command. When the stress response is triggered—whether by a looming deadline, chronic inflammation, emotional trauma, or poor sleep—the baton is passed to cortisol, and every other hormone bows in submission.

At the top of this command structure sits the HPA axis—the hypothalamic-pituitary-adrenal axis. While the name might sound clinical, its effects are anything but abstract. This axis governs the way your body interprets and responds to stress. Think of the hypothalamus as the general in a control tower scanning the environment for any signs of threat. When it detects even a whisper of danger—real or imagined—it signals the pituitary, which acts as a relay officer. The pituitary releases ACTH (adrenocorticotropic hormone), which tells the adrenal glands to flood your system with cortisol.

Cortisol, often labeled the "stress hormone," is not inherently bad. It's lifesaving in acute danger. It mobilizes energy, heightens awareness, and suppresses non-essential functions like reproduction and digestion, which aren't immediately needed when fleeing from a predator or navigating a crisis. The problem is, our modern lives have redefined what qualifies as a "crisis." Instead of intermittent, short bursts of stress followed by recovery, we now experience a slow, simmering boil—constant emails, emotional overwhelm, blood sugar crashes, overexposure to artificial light, underexposure to nature, and chronic sleep deprivation. This relentless state of perceived threat forces cortisol to remain elevated, and once that happens, the hormonal hierarchy begins to collapse.

High cortisol levels have a ripple effect across nearly every hormonal system. One of the most critical consequences of chronic stress is what's known in functional medicine as "pregnenolone steal." Pregnenolone is often referred to as the "mother hormone"—it's the raw material from which your body produces other hormones, including DHEA, progesterone, estrogen, testosterone, and cortisol. When you're in a chronic stress state, the body prioritizes survival over reproduction and long-term repair. To keep cortisol levels high, it diverts pregnenolone away from the sex hormone pathways and funnels it into the cortisol production line. This isn't an error in design— it's a brilliant evolutionary adaptation meant for short-term stress. But in the long term, it leaves the body depleted.

The symptoms of this hormonal redirection are wide-ranging and easy to dismiss until they become chronic. Women may experience irregular cycles, low libido, mood swings, or infertility. Men may feel fatigue, low motivation, diminished performance, and emotional flatness. Both sexes may struggle with anxiety, weight gain around the midsection, insomnia, and poor recovery from exercise. These aren't simply the byproducts of aging or a busy schedule—they're signals that the endocrine hierarchy is being hijacked by an outdated stress protocol.

What's often misunderstood is that hormones don't operate in isolation. They're part of a dynamic feedback loop that functions in accordance with circadian rhythms—biological cycles wired into us through millions of years of evolution. Cortisol, for instance, is meant to rise early in the morning, peaking around sunrise to help you wake up, mobilize energy, and start the day. As the day progresses, cortisol should gently taper off, allowing melatonin—your sleep hormone—to rise in the evening. This interplay between cortisol and melatonin orchestrates your sleep-wake cycle, governs when you eat, how you digest, and even when you repair cellular damage.

When this natural rhythm is disrupted—say, by checking your phone at 2:00 AM or staying under blue light until midnight—the entire system is thrown off. Melatonin suppression leads to poor sleep quality, which exacerbates cortisol elevation the next morning, leading to insulin resistance and poor glucose handling throughout the day. You crave sugar and caffeine not because you're weak, but because your hormones are attempting to self-regulate under duress. You eat late at night, and your digestive enzymes underperform. The food lingers longer in your stomach, creating bloating, discomfort, and inflammation. Meanwhile, growth hormone—the key hormone responsible for cellular repair, lean muscle maintenance, and fat burning—is being suppressed because your sleep is fragmented. And just like that, a single circadian misalignment has thrown off half a dozen major hormonal systems.

Even thyroid function—often blamed when people feel tired or gain weight—can be secondary to stress. Cortisol downregulates the conversion of inactive T4 to active T3, which is the form your cells actually use. High cortisol also increases reverse T3, a "decoy" version of thyroid hormone that binds to receptors but does nothing, essentially jamming the signal. So, you could have a thyroid panel that looks "normal" on paper, but functionally you're in a hypothyroid state because cortisol is interfering with signal transmission. The result is fatigue, sluggish metabolism, cold intolerance, and difficulty concentrating.

Sex hormones fare no better. Estrogen and testosterone are especially sensitive to cortisol's dominance. In men, chronically elevated cortisol lowers testosterone production and impairs the Leydig cells in the testes that manufacture it. In women, cortisol throws off the delicate balance of estrogen and progesterone, often leading to estrogen dominance, which can manifest as PMS, mood swings, fluid retention, and fibrocystic breasts. Over time, the body begins to interpret stress as the default state and turns down reproductive capacity altogether. It's not uncommon to see younger and younger people facing infertility not because of mechanical issues, but because their internal signal says, "This environment isn't safe to reproduce."

The endocrine system doesn't just stop at sex hormones and thyroid. Insulin, the hormone responsible for regulating blood sugar, is closely tied to cortisol as well. Cortisol raises blood glucose in preparation for action, but if that action doesn't come—say, you're sitting in traffic or at your desk—the glucose has nowhere to go. Over time, this leads to insulin resistance, where cells become desensitized to insulin's signal. This isn't just a blood sugar issue; it's a hormonal issue. Insulin resistance is tied to increased inflammation, impaired fat metabolism, and accelerated aging. The link between stress and metabolic syndrome becomes undeniable once you view it through the lens of hormonal hierarchy.

And then there's growth hormone, the unsung hero of tissue repair, neurogenesis, fat metabolism, and longevity. It is primarily released during deep, slow-wave sleep—a phase of the sleep cycle that is highly sensitive to cortisol. Elevated nighttime cortisol blunts growth hormone secretion, impairing the body's ability to heal, build muscle, and detoxify. The visible signs include slow wound healing, loss of muscle mass, and increased fat storage. Internally, the consequences are even more profound—reduced neuroplasticity, poor immune resilience, and faster cellular aging.

What emerges is a clear and sobering picture. Our hormonal systems are not malfunctioning because of random bad luck or genetic fate. They are responding precisely as they were designed to respond—to survive a world that feels hostile, dangerous, or unpredictable. Unfortunately, that "hostile world" now includes everything from processed food to artificial light, sedentary habits, unresolved trauma, and digital overload. The stressors are constant, the recovery windows are short, and the body adapts accordingly, often at the cost of balance and well-being.

Understanding this hierarchy—truly grasping how and why your hormones respond the way they do—is the first step in reclaiming your health. It shifts the conversation from isolated symptoms and fragmented treatments to systemic root causes. Fatigue is no longer just about low thyroid. Weight gain is not merely about willpower. Infertility isn't always a reproductive problem. These are all expressions of a deeper imbalance where the HPA axis has seized control, and the rest of the hormonal orchestra has been forced to follow a chaotic, dissonant tune.

The good news is that the endocrine symphony can be retuned. But that requires more than supplementation or pharmaceutical interventions. It demands a lifestyle shift that communicates safety, rhythm, and nourishment to your body—environments that signal to your ancient biology that it's okay to release the gas pedal and engage healing, repair, and growth.

In the coming sections, we'll explore how to create those signals intentionally—through light, sleep, nutrition, movement, and nervous system regulation. Because when cortisol relinquishes its grip on the hormonal throne, the entire endocrine hierarchy begins to harmonize once again. And in that harmony, vitality returns.

7.2 Modern Endocrine Disruptors and Hormonal Chaos

Modern life does not simply challenge the body with emotional and psychological stress. It bombards the endocrine system with environmental toxins and invisible influences that mimic or disrupt natural hormonal rhythms. Among the most insidious of these are xenoestrogens—foreign compounds that behave like estrogen in the body. These synthetic substances are found in plastics, pesticides, cosmetics, cleaning agents, and even in the lining of food cans. They do not merely enter and exit the body unnoticed. They bind to estrogen receptors, hijack cellular signaling, and alter gene expression in ways that can confuse the body's natural hormonal balance.

Xenoestrogens contribute to a state of estrogen dominance, even in the absence of high endogenous estrogen production. Women may experience heavy periods, fibroids, endometriosis, and breast tenderness. Men, meanwhile, may face reduced testosterone levels, gynecomastia, and lower sperm counts. This hormonal confusion is particularly dangerous during critical windows of development—such as fetal growth, puberty, and pregnancy—when endocrine systems are more sensitive and malleable. The rise in hormone-sensitive cancers, including breast, ovarian, prostate, and testicular cancer, has been linked in part to long-term exposure to these environmental mimics.

Compounding the problem is the increasing exposure to electromagnetic fields (EMFs) from wireless technology. While the convenience of mobile phones, Wi-Fi, and wearable tech is undeniable, research has begun to unveil their subtle but pervasive impact on hormonal function. The pineal gland, a tiny endocrine organ responsible for melatonin production, is particularly sensitive to EMFs. Disruption to melatonin secretion doesn't just impair sleep—it influences the entire hormonal cascade, since melatonin interacts with the HPA axis, reproductive hormones, and immune signaling.

In both men and women, chronic exposure to non-ionizing radiation has been correlated with altered levels of luteinizing hormone, follicle-stimulating hormone, and testosterone. Animal studies have shown reduced fertility rates, delayed puberty, and abnormal reproductive anatomy when exposed to EMF sources during key developmental stages. While the full implications for humans are still being studied, the early evidence suggests a biological impact that cannot be ignored.

Overlaying all of this is the relentless drumbeat of chronic stress, which continues to rewrite hormonal scripts across populations. Persistent elevation of cortisol does more than steal from other hormones—it reshapes metabolic priorities. One of the first systems to be affected is insulin regulation. When cortisol remains high, glucose remains elevated. Insulin is released to shuttle the excess sugar into cells, but over time, the cells become resistant to its signal. This insulin resistance is the gateway to metabolic syndrome, type 2 diabetes, and fat accumulation— particularly visceral fat, which itself is hormonally active and inflammatory.

Leptin, the hormone that signals satiety, is also thrown off balance. Chronically elevated insulin can lead to leptin resistance, in which the brain no longer receives the signal to stop eating. The result is a vicious cycle of overeating, weight gain, and further hormonal disruption. Thyroid function also suffers under chronic stress. As mentioned earlier, cortisol inhibits the conversion of T4 to T3, while promoting the production of reverse T3, which further blocks thyroid receptor activation. Thus, stress not only makes people tired—it slows their metabolism, promotes fat gain, and suppresses their ability to lose weight.

In sum, modern endocrine disruptors—from synthetic chemicals to radiation to emotional stress—are not working in isolation. They converge on a shared pathway: the dysregulation of the body's natural hormonal intelligence. This convergence is why so many people feel like they are doing "everything right" yet still experience weight gain, fatigue, mood swings, and reproductive struggles. Until the root environment is addressed, symptoms will persist, and true healing will remain elusive.

7.3 Restoring Hormonal Balance Through Root Cause Medicine

The answer to this hormonal chaos is not found in chasing symptoms or patching the leaks with isolated hormone replacements. True restoration begins by addressing the upstream causes of imbalance and creating an internal environment where the endocrine system can remember its original harmony. The first and most fundamental step is managing stress—not in the superficial sense of "relaxation," but in cultivating real physiological safety.

When the body perceives safety, cortisol production diminishes, and the HPA axis gradually returns to its natural rhythm. To support this, adaptogenic herbs such as ashwagandha, rhodiola, and holy basil have been shown to modulate the stress response without blunting its necessary spikes. These plants help buffer the impact of acute stress while retraining the system to respond proportionately. Paired with this botanical support, mind-body practices such as meditation, breathwork, forest bathing, and restorative movement communicate to the brain that it is safe to downregulate the stress response.

Another vital piece of the puzzle is the liver—the body's primary detoxification organ and one of the most important regulators of hormone metabolism. Hormones, especially estrogen, must be properly broken down and excreted through a two-phase process in the liver. In phase I, hormones are made water-soluble. In phase II, they are bound to specific compounds that allow for safe elimination via the bile or urine. If either of these phases is sluggish or overwhelmed—due to poor diet, alcohol, medications, or environmental toxin load—hormonal metabolites can recirculate and cause damage.

Supporting liver detoxification involves both removing the burden and enhancing function. Reducing alcohol, processed foods, and chemical exposure lowers the load, while nutrients like B vitamins, magnesium, selenium, and amino acids such as glycine and taurine fuel the detox pathways. Bitter herbs like dandelion root, burdock, and milk thistle can stimulate bile flow, while cruciferous vegetables like broccoli and cauliflower provide indole-3-carbinol, which supports estrogen metabolism through favorable pathways.

Only once stress regulation and detoxification are in place does it make sense to consider targeted hormone support. In cases of significant hormonal depletion, bioidentical hormone replacement therapy (BHRT) can offer relief and restoration. However, this must be done carefully and under the guidance of professionals who understand endocrine feedback loops. Simply adding estrogen, progesterone, or testosterone to a system in chaos may produce short-term benefits but long-term complications if the root dysfunction persists.

In some cases, the use of hormone precursors such as DHEA or pregnenolone may gently nudge the body toward balance without overriding its own production. Herbal hormone modulators like chasteberry, maca, and tribulus can also be used to support the body's internal hormone production and receptor sensitivity. The goal is not to force hormone levels into an artificial norm, but to restore the body's capacity to self-regulate.

What emerges from this approach is a reimagining of hormonal health. No longer is it a game of numbers and lab ranges, but a reflection of how well the body is perceiving, processing, and responding to its environment. Hormones are messengers. When they are out of balance, they are not the problem—they are the signal. To restore hormonal balance is to listen to the message and transform the terrain. It is to create a life that whispers safety to the nervous system, nourishment to the cells, and rhythm to the clock within. Only then can the endocrine symphony return to playing in tune—and when it does, the entire organism remembers what it means to be alive, aligned, and fully human.

Chapter 8: The Forgotten Senses – How Sensory Deprivation Is Making You Sick

8.1 The Sensory Systems Modern Life Has Hijacked

We tend to think of our senses as passive tools, windows through which we perceive the world without consequence. But what if the reverse is also true? What if the way we shape our environments — the sounds we filter out, the light we surround ourselves with, the surfaces we walk on, the electromagnetic frequencies we bathe in — is not just influencing our perception, but actively degrading our health?

We evolved immersed in sensory richness. Our ancestors moved across varied terrain, barefoot or lightly shod, navigating forests, sand, stones, uneven earth. They rose with the sun and fell asleep to darkness, attuned to the gradual shift in light throughout the day and seasons. They bathed in natural electromagnetic fields, breathed unprocessed air rich in information, and constantly calibrated their bodies in response to texture, vibration, heat, and sound. In contrast, modern humans live in what amounts to a low-grade sensory deprivation chamber. We spend our days on smooth floors, beneath flickering blue LEDs, in climate-controlled buildings, eyes fixed on screens, feet caged in shoes, nervous systems locked in artificial rhythms that the body interprets not as safety — but as a subtle, chronic threat.

Let's begin with proprioception, the sense of bodily awareness and orientation in space. Most people rarely think about it — until they begin to lose it. It's proprioception that tells you where your limbs are without needing to look, that lets you walk in the dark without tripping, that allows a gymnast to flip, a dancer to spin, a child to climb. This deeply embodied intelligence is trained through interaction with the real world — through diverse movement, through instability, through sensation. But in modern life, our movement patterns have become absurdly restricted. We walk on flat, paved surfaces. We sit for hours in the same positions, wearing shoes that dull feedback and eliminate ankle flexion. Over time, proprioceptive input withers. The brain begins to lose the high-resolution map it once had of the body. Injuries increase. Balance falters. Coordination suffers. And with that comes a growing disconnection not just from the world, but from ourselves.

Loss of proprioception doesn't happen in isolation. It is often accompanied by vestibular dulling — the sense of balance and acceleration that relies on motion. Humans were not designed to live in boxes, to drive instead of run, to ride elevators instead of climb. Our inner ears were meant to be challenged — by jumping, spinning, climbing, hanging, rolling. Remove those movements and the brain loses calibration. Dizziness, nausea, car sickness, and poor balance in old age are not random afflictions. They are the logical consequences of a life deprived of sensory richness.

But proprioception and movement are only one domain. The assault continues through our light environment. For most of human history, light came from the sun, fire, moon, and stars. This light was dynamic, changing across time and space, with full spectrums of ultraviolet, infrared, and visible wavelengths. Modern lighting, however, is profoundly unnatural. LED bulbs, fluorescent tubes, and digital screens emit narrow bands of light — most notably, excessive blue wavelengths and almost no infrared. Blue light isn't inherently bad. During the day, it plays a vital role in regulating circadian rhythms, keeping us alert and engaged. But when we are exposed to artificial blue light late into the night — often inches from our eyes via smartphones and laptops — our melatonin production is suppressed, our sleep cycles disturbed, and the entire hormonal cascade that depends on the light-dark cycle begins to unravel.

Natural light contains far more than meets the eye. Infrared light, especially from the early morning sun, stimulates cytochrome c oxidase in the mitochondria, enhancing ATP production and cellular energy. UVB light allows the skin to manufacture vitamin D, which in turn regulates calcium metabolism, immune function, and over 1000 genes. When we live under fake light, rarely go outside, or slather on sunscreen every time we do, we are not simply avoiding sunburn — we are depriving ourselves of fundamental data that the body needs to function.

Our electromagnetic environment is another invisible layer of disruption. Earth emits a natural electromagnetic frequency known as the Schumann resonance, roughly 7.83 Hz. This low-frequency pulse arises from lightning strikes and resonates within the space between the Earth's surface and the ionosphere. Remarkably, this frequency closely matches alpha brainwaves — the state associated with calm wakefulness, meditative flow, and creativity. For most of human evolution, our biology evolved under the constant influence of this resonance. But in the last century, the electromagnetic landscape has changed more than in the previous thousand. We now live amidst an explosion of high-frequency artificial EMFs — from cell towers, Wi-Fi routers, smart meters, Bluetooth devices, and 5G infrastructure.

These man-made frequencies, unlike the Schumann resonance, are pulsed and non-harmonious. They do not simply add to the electromagnetic soup — they replace it. The implications are staggering. Emerging research has linked chronic EMF exposure to a range of physiological disruptions: reduced melatonin secretion, increased oxidative stress, blood-brain barrier permeability, altered calcium signaling, and mitochondrial dysfunction. While some individuals suffer acutely from electromagnetic hypersensitivity, experiencing headaches, insomnia, anxiety, or heart palpitations near EMF sources, many more are likely experiencing subtle, cumulative effects they do not consciously associate with their environment. As with poor lighting or sedentary movement, the body may not scream — but it whispers, and over time, the whispers become pathology.

The artificial environments we've constructed don't just fail to stimulate our senses — they confuse them. Consider texture and temperature. Modern buildings are designed for efficiency and uniformity. Surfaces are flat and predictable. Rooms are held at constant temperatures. We wear shoes on carpets, sit on molded chairs, touch cold glass screens. Compare this to the complex textures of bark, moss, gravel, sand, fur, skin. Or the thermal variability of a forest trail — sun and shadow, cool stream, warm stone. These textures are not just sensory delights; they're data. They feed the nervous system information that helps calibrate everything from grip strength to thermoregulation.

When these textures are removed, the nervous system is starved. Children raised in sterile, overly controlled environments often exhibit sensory processing difficulties — hypersensitivity to touch, clumsiness, inability to regulate arousal. These aren't personality quirks; they are symptoms of developmental deprivation. Adults suffer too, but often label their symptoms differently: burnout, fatigue, mood swings, loss of joy. Beneath the labels lies a common thread — the nervous system is overwhelmed by artificial inputs and undernourished by natural ones.

This deprivation even extends to olfaction. The modern world is full of synthetic fragrances — detergents, air fresheners, perfumes, candles — that overload the olfactory system and dull its sensitivity. Natural smells like rain, pine, soil, and ocean carry microbial and chemical signatures that interact with immune function, memory, and even the endocrine system. Studies show that exposure to natural scents reduces cortisol levels, enhances parasympathetic tone, and improves cognitive performance. Yet, the average urban dweller spends 90% of their time indoors, breathing recycled air and volatile organic compounds instead.

The tragedy is that most people don't realize what they've lost, because they've never experienced it. They adapt to low-grade discomfort — poor sleep, low energy, brain fog, irritability — and assume it's normal. But this is not normal. It is the outcome of living in a world that has forgotten what it means to be human. To heal, we must remember. We must restore the inputs our biology evolved to expect.

The sensory hijacking of modern life isn't merely about discomfort. It is about disorientation at the most fundamental level. The nervous system, deprived of its expected signals — real light, natural EMFs, rich textures, dynamic movement — goes into hypervigilance. It perceives the mismatch as a threat. This is not imagination. It is physiology. And it's happening silently in millions of bodies around the world.

Reconnection doesn't require abandoning civilization. It requires intention. It means walking barefoot on the earth, watching the sunrise without sunglasses, turning off Wi-Fi at night, lighting candles instead of bulbs, touching wood and stone, listening to birdsong instead of traffic. It means reclaiming the full spectrum of sensory life, not as a luxury, but as a biological imperative. Because without it, the body forgets how to regulate, how to relax, how to thrive. And in that forgetting, disease takes root.

In the next section, we'll explore how to restore and protect your sensory systems, not with supplements or gadgets, but through deliberate re-engagement with the natural world — the world your senses have always known how to love.

8.2 Sensory Deprivation and Its Health Consequences

While much of modern medicine focuses on chemical imbalances and genetic predispositions, it often overlooks a far more ancient and immediate factor in human health: the quality and quantity of our sensory environment. We are not minds floating in isolation — we are embodied beings whose survival and vitality depend on the continuous flow of sensory data from the external world. When that flow is disrupted, narrowed, or hijacked, the consequences are not only psychological but deeply physiological. Among the most insidious effects of sensory deprivation is the body's slow withdrawal from its own healing intelligence.

Consider the widespread crisis of vitamin D deficiency. Millions of people in developed countries are supplementing with vitamin D3, yet blood levels remain suboptimal, immune systems remain compromised, and the expected benefits — better bone health, mood stability, and hormonal regulation — often fail to manifest. The reason lies not only in dosage or absorption, but in the form and origin of the vitamin itself. True vitamin D production happens on the skin, in direct response to UVB rays from sunlight. This process doesn't merely generate cholecalciferol; it initiates a cascade of photo-biochemical events that yield sulfated vitamin D3 — a water-soluble, transport-ready version that can circulate freely in the bloodstream, unlike the fat-soluble oral form.

This sulfated version appears to behave differently in the body, interacting with cholesterol sulfate on the skin and supporting the structure and electrical charge of red blood cells, among other things. It's not simply a vitamin; it's part of a larger electrochemical system. Without sunlight, this system becomes impaired. No supplement can fully replicate the hormonal, electromagnetic, and enzymatic nuances of real sunlight on bare skin. And yet, most people spend over 90% of their lives indoors, under artificial lighting that lacks UVB entirely. Worse, when they do go outside, they're often covered in clothing or sunscreen, further blocking the necessary frequencies. The result is a society both supplement-rich and sunlight-starved — a paradox that speaks to the limits of pharmaceutical thinking in the face of sensory deprivation.

Another consequence of our environmental disconnection is the depletion of electrical grounding. The Earth's surface carries a subtle negative charge due to a near-constant flow of electrons, a charge maintained by lightning strikes and atmospheric ionization. Our ancestors, walking barefoot, sleeping on the ground, and touching trees, were constantly absorbing this free source of electrons. These electrons are not abstract: they neutralize reactive oxygen species, reduce inflammation, and help stabilize the electrical potential of every cell in the body. Today, we live on the second floor, walk on rubber soles, and sleep on insulated mattresses. The skin rarely touches the Earth.

This loss of connection leads to a kind of low-grade charge starvation. Chronic inflammation, poor wound healing, increased oxidative stress — all correlate strongly with the absence of grounding. It's not that humans are deficient in anti-inflammatory drugs; they're deficient in electrons. And this deficiency, while invisible, contributes to the rising tide of autoimmune conditions, chronic fatigue, and metabolic disorders. When people do reconnect — through barefoot contact with grass, sand, or conductive devices — physiological shifts are measurable within minutes: cortisol levels normalize, heart rate variability improves, and blood viscosity decreases. The body is not designed to be electrically isolated from its source. To do so is to invite dysfunction at every level.

Sound is another vital sense that modernity has hijacked. The human auditory system evolved in a world of organic acoustics:

rustling leaves, bird calls, flowing water, the distant roll of thunder. These natural frequencies follow fractal patterns and harmonic ratios that are inherently calming to the nervous system. In contrast, artificial sounds — alarms, traffic, air conditioners, computer fans, electronic notifications — are often abrupt, monotonous, and disharmonious. They lack the richness and variability that natural sounds provide, and they activate the stress response far more easily. The auditory system, when bombarded with such mechanical noise, adapts by reducing sensitivity, altering frequency discrimination, and triggering chronic stress pathways in the limbic system.

Research on sound therapy and nature exposure confirms what intuition has long suggested. Natural sounds, especially those that mimic flowing water or birdsong, lower cortisol, improve parasympathetic tone, and enhance cognitive performance. They do not simply entertain — they entrain. They signal safety. They calm the amygdala. They remind the body that it is still embedded in a larger system of rhythm and coherence. Without these sounds, the nervous system becomes brittle. Hypervigilance sets in. Sleep suffers. Emotional regulation declines. The modern soundscape is not neutral — it is a physiological stressor, and its impact is cumulative.

In sum, the deprivation of light, touch, sound, movement, and electromagnetic contact isn't just a cultural loss — it is a biological threat. The body interprets this deprivation as a sign that something is wrong, that the environment is unsafe, incomplete, or dying. This interpretation isn't metaphorical; it is biochemical. Hormones shift. Neurotransmitters adapt. Organs adjust their outputs. Over time, the result is not just illness, but confusion — a body that no longer knows how to orient itself in the world because the signals it evolved to interpret have gone silent.

Reversing these effects doesn't start with more medication. It starts with restoring the missing sensory inputs — reintroducing the forgotten languages that the body once spoke fluently.

8.3 Sensory Restoration Protocols for Modern Humans

If sensory deprivation can cause illness, then sensory restoration is not simply therapeutic — it is foundational. The modern body, bombarded by artificiality, craves a return to the real. But that return doesn't require a monastic retreat into the forest. It can begin, powerfully, with simple, deliberate acts that reintroduce the body to the elements it was designed for. The process is not merely psychological or recreational — it is physiological rewilding.

Start with the feet. There is no more immediate or effective way to restore connection with the Earth than to remove the shoes and walk on natural surfaces. Barefoot walking engages thousands of nerve endings, strengthens stabilizer muscles, activates dormant proprioceptive pathways, and — most importantly — facilitates direct electron transfer from the ground into the body. Even ten minutes of barefoot time per day has been shown to improve mood, reduce inflammatory markers, and enhance sleep quality. For those in urban environments, grounding mats and conductive bed sheets can serve as proxies, restoring the body's electrical relationship with the planet, particularly during sleep, when recovery processes are most active.

Light restoration is equally critical. The first light the body sees in the morning sets the tone for every hormonal rhythm that follows. Morning exposure to natural sunlight — especially within the first hour after waking — triggers a surge of cortisol that is healthy and necessary. This peak tells the brain that it is time to be alert, and it anchors the timing of melatonin release that night. Ideally, this exposure should last at least 10 to 15 minutes, with no sunglasses or windows between you and the sun. Midday sun, especially in seasons with higher solar angles, provides critical UVB for vitamin D synthesis, while infrared rays nourish mitochondria and improve circulation.

Artificial light management is equally important. After sunset, lighting should shift to the red and amber ends of the spectrum. This can be done through incandescent bulbs, red LED lights, candlelight, or blue-blocking glasses. Screen exposure should be minimized in the two hours before sleep, or at the very least filtered through software that adjusts color temperature. These shifts help the pineal gland resume melatonin production and allow the endocrine system to follow its ancient rhythm — light means action, darkness means repair.

Touch and texture can be reintroduced gradually and joyfully. Walk barefoot not only on grass, but on stone, wood, sand, and soil. Let your skin feel wind, water, rain. Sit on the ground. Touch bark, leaves, fur, moss. Vary your environment. Each new texture feeds the brain, enlivens dormant pathways, and reinforces the sense of aliveness. Likewise, cold exposure — through cold showers, ocean swims, or river dips — stimulates norepinephrine release, sharpens mental clarity, reduces inflammation, and trains the vascular system to adapt more efficiently to thermal stress.

Movement should also be varied and multi-planar. Repetitive gym routines do not replace the rich complexity of climbing, crawling, twisting, balancing, and exploring. Restore play. Rediscover nonlinear motion. Even simply walking in nature — navigating roots, adjusting to inclines, reacting to birds or shifting light — reawakens the body's need to respond to novelty. This novelty is not chaos; it is coherence. It trains resilience.

Finally, enrich your soundscape. Play recordings of nature if you're in the city — not as background noise, but as intentional listening sessions. Better yet, seek out real natural sound whenever possible. Sit beside water, even if it's a fountain. Walk under trees filled with birds. Learn to identify their calls. Close your eyes and let your nervous system recalibrate to frequencies that do not demand anything from you but awareness.

Restoring sensory intelligence is not about escapism — it is about returning to a conversation that your body never wanted to stop having. Every nerve ending is a tongue. Every cell is an ear. Every breath is a message. Health is not the absence of disease — it is the full expression of your capacity to relate to the world around you. And in a world designed to dull that relationship, choosing to re-engage becomes a radical act of healing.

Chapter 9: Quantum Biology – The Energy Fields That Control Your Health

9.1 Quantum Mechanics in Living Systems

The world you see, touch, and measure is not the one that truly governs life. Beneath the mechanical gears of molecules and the elegant logic of cellular pathways lies a subtler domain — a realm where particles can be in two places at once, where effects can ripple backward through time, and where entangled elements mirror each other across space without any physical link. This is the world of quantum mechanics — and astonishingly, it is not confined to particle accelerators or physics textbooks. It is the very language of life itself.

For decades, biology dismissed quantum physics as irrelevant, a theory for physicists and cosmologists to ponder, not physicians or biochemists. The inner workings of the body were mapped through Newtonian eyes: predictable reactions, local causes, mechanical responses. But in recent years, a quiet revolution has taken hold. With the help of ultra-sensitive imaging tools and increasingly precise molecular models, researchers are uncovering evidence that living organisms — from single-celled bacteria to the complex architecture of the human brain — are not just influenced by quantum mechanics but utterly dependent on it.

At the heart of this realization is **quantum coherence**, a phenomenon that allows particles — including electrons and protons — to exist in multiple states simultaneously. In traditional chemistry, a reaction proceeds one step at a time, like moving a marble through a maze. But quantum coherence offers another route entirely: particles can "explore" all possible paths at once and choose the most efficient one. In the biological world, this is not science fiction. This is how **enzymes**, those molecular machines that control virtually every reaction in your body, operate.

Take, for example, **enzyme catalysis**. Enzymes accelerate chemical reactions by factors of millions or billions, enabling life processes that would otherwise take centuries to occur. Classical theories could not fully explain this speed. The missing piece turned out to be quantum tunneling — the ability of a particle to pass through energy barriers instead of climbing over them. In the case of enzyme reactions, hydrogen atoms tunnel through activation energy barriers, making biochemical transformations nearly instantaneous. This is not an exception; it's a feature. Quantum tunneling is built into the very architecture of the enzyme's active site. In effect, the enzyme doesn't just nudge a molecule into reacting — it creates the perfect quantum landscape for a miracle to happen.

Even photosynthesis, the process that powers nearly all life on Earth, depends on quantum effects. When a photon of sunlight hits a leaf, it excites an electron within the plant's light-harvesting complex. That electron doesn't simply move randomly through the system until it finds the right chemical partner. Instead, it travels along all possible paths simultaneously — a quantum process known as superposition — and finds the most efficient route in femtoseconds. Without this quantum efficiency, photosynthesis would be far too slow to sustain life. The very act of converting light into usable energy — a fundamental process for every food chain — hinges on principles that defy classical understanding.

But perhaps the most controversial and thrilling application of quantum biology lies in the human brain. For years, consciousness has remained a scientific mystery, untouched by advances in neuroscience. While we've mapped brain regions, identified neurotransmitters, and developed intricate models of cognition, we've never explained how subjective experience — awareness, thought, memory, emotion — arises from the physical machinery of the brain. Enter the theory of **microtubule quantum processing**.

Microtubules are cylindrical protein structures found in nearly every cell, but especially dense in neurons. Traditionally, they've been viewed as scaffolding — a structural support for cellular shape and division. But researchers like Sir Roger Penrose and anesthesiologist Stuart Hameroff have proposed something far more radical: that microtubules function as quantum computers. Unlike conventional processors, which use binary logic (on/off), quantum computers rely on qubits — units of information that can exist in superpositions, allowing massively parallel calculations.

Inside neurons, microtubules may act as information processors on a quantum level, enabling consciousness to arise not from brute electrochemical firing, but from orchestrated quantum events. The brain, in this view, is not just a chemical soup but a finely-tuned quantum instrument. It perceives, stores, and reacts to information in ways that exceed the limits of classical computation. Moreover, the coherence of quantum states in microtubules — maintained perhaps through specific environmental conditions in the brain — could explain the speed and integration of thought, the fluidity of memory, even the phenomenon of intuition.

Skeptics have rightly questioned how such delicate quantum effects could persist in the warm, wet environment of the human body, which would typically cause decoherence — the collapse of quantum states — in nanoseconds. Yet evidence is emerging that biological systems have evolved ingenious ways of shielding or sustaining coherence. Structures like enzyme pockets, photosynthetic membranes, and potentially microtubules create quantum-friendly environments, where noise is reduced and order is maintained long enough for these phenomena to occur. Life, in other words, may be nature's way of protecting and exploiting quantum mechanics.

Another extraordinary aspect of quantum biology is the behavior of **DNA as an electromagnetic antenna**. The double helix of DNA is more than a storage device for genetic information; its spiral geometry, electrical properties, and resonant frequency make it an active participant in the body's quantum communication system. Studies suggest that DNA can emit and receive electromagnetic signals in specific frequencies — a concept known as **bio-resonance**. This suggests that gene expression may not be controlled solely by chemical signals like transcription factors, but also by energy fields — subtle, frequency-specific patterns that influence which genes are switched on or off.

The implications are enormous. It means that your genetic expression — and thus your susceptibility to disease, your capacity for repair, your emotional tendencies — may be shaped not only by what molecules are present in your cells, but by the energetic environment they inhabit. Light, sound, electromagnetic frequencies, and even intention could influence biological outcomes via quantum resonance. This offers a powerful complement to the traditional gene-centric model of biology. You are not merely a passive recipient of DNA instructions; your internal and external environments are co-authors in the script.

And while these ideas challenge deeply-held assumptions in Western medicine, they are not entirely new. Ancient healing systems — from Chinese medicine to Ayurveda — have long spoken of life force, energy channels, and vibrational health. What quantum biology does is provide a scientific lens through which to reinterpret these insights. The "meridians" of acupuncture may correspond to conductive pathways of bioelectrical flow. The chakras may align with plexuses of neuroendocrine activity that emit and respond to electromagnetic fields. What once seemed mystical now begins to look like pre-modern models of quantum systems — just expressed in a different language.

Still, the most radical shift quantum biology demands is not technological, but philosophical. If the body is quantum, then it is not a machine to be repaired in pieces. It is a field — dynamic, interconnected, and exquisitely sensitive to context. Healing, in this view, involves more than correcting chemical imbalances or removing dysfunctional tissues. It requires coherence. It requires resonance. It requires restoring the subtle energetic architecture that underlies health.

It also means that human beings are not closed systems. You are not limited by your physical boundaries. Your consciousness, your cells, even your genetic code interact continuously with the environment in non-linear, non-local ways. The implications for health are profound. It may not be enough to ask what's wrong with a person — we must ask what frequencies they're immersed in, what rhythms they've lost, what coherence has been disrupted.

Quantum biology does not reject classical medicine; it extends it. It recognizes that molecules matter, but they are not the whole story. It honors the role of the physical while reclaiming the reality of the energetic. And it invites us to reimagine the body not as a biochemical robot, but as a quantum symphony — a living orchestra of light, energy, and vibration, tuned moment by moment to the frequencies of life itself.

9.2 Electromagnetic Biology and Health

For decades, the conversation around health has revolved almost exclusively around chemical pathways and biochemical interactions. Yet, there is a deeper layer of communication within the human body that precedes chemistry: electromagnetism. Every cell, every neuron, every organ emits and responds to subtle electromagnetic fields. The heart alone produces an electromagnetic field measurable several feet from the body. The brain, with its intricate electrical impulses, generates frequency-based rhythms that synchronize with bodily states. But what happens when these natural electromagnetic rhythms are continuously bombarded by artificial signals?

Modern technology has cloaked our environment in a dense web of man-made electromagnetic frequencies. Cell phones, Wi-Fi routers, Bluetooth signals, power lines, and satellite transmissions create a 24/7 cascade of electromagnetic noise. These signals operate in ranges that are foreign to biological systems, particularly in their intensity, modulation, and constant presence. The human organism did not evolve within such an environment. As a result, it interprets this bombardment as a form of chronic environmental stress.

Artificial electromagnetic fields interfere with the body's native bioelectrical signals in several key ways. At the cellular level, EMFs have been shown to affect voltage-gated calcium channels—membrane proteins that regulate the influx of calcium ions into the cell. Excess intracellular calcium can create oxidative stress, DNA strand breaks, and mitochondrial dysfunction. This is not fringe speculation. Numerous peer-reviewed studies now show that prolonged exposure to EMFs can reduce sperm motility, lower melatonin levels, and even promote the formation of aggressive cancers in animal models. And these are not high-intensity exposures. Even low-level, long-duration contact has been implicated in disrupting cellular signaling pathways.

Beyond calcium channels, EMFs affect the body's overall energetic coherence. Bioelectric communication is how cells talk to one another, coordinate healing, and regulate function. Disrupting this flow is akin to introducing static into a radio signal. The message may still come through, but it's garbled, delayed, or entirely misinterpreted. This interference is especially troubling for tissues that rely heavily on electrical precision, such as cardiac muscle, neuronal networks, and endocrine glands.

Parallel to these disturbances, emerging research has brought new attention to water—not just as a solvent in the body, but as an information carrier. Water inside the human body is not inert or structureless. Recent findings suggest that water near hydrophilic surfaces—such as cell membranes—forms coherent domains, layers of structured water that behave more like liquid crystals than traditional fluid. This structured water appears to store and transmit information through quantum coherence. In other words, it becomes a conduit for electromagnetic signaling within and between cells.

Structured or "coherent" water plays a key role in maintaining the cell's electrical potential. Cells are essentially tiny batteries, maintaining a voltage gradient across their membranes. Structured water helps to hold and stabilize this voltage, improving the efficiency of electron transport and enzyme function. Disruption of this coherence through EMF exposure or dehydration may explain why so many chronic illnesses are marked by low energy, brain fog, and immune disarray. You're not just thirsty—you're electrically unstable.

Then there's the realm of frequency-specific healing, a domain long marginalized by conventional science but now returning under the lens of quantum biology. Ancient cultures have long recognized the power of sound and vibration in healing. Tibetan singing bowls, Gregorian chants, and indigenous drumming are not random artistic expressions—they are bioenergetic tools that alter the human state through vibrational entrainment. What modern science is beginning to observe is that specific frequencies do indeed produce measurable biological effects.

For example, the frequency of 528 Hz has been studied for its potential in DNA repair and cellular regeneration. In controlled environments, this frequency has been observed to increase UV light absorption in DNA samples, which may be indicative of a structural or energetic shift in the molecule. Another frequency, 40 Hz, aligns with gamma brainwaves, associated with heightened states of awareness, cognitive function, and even neuroplasticity. Alzheimer's studies have shown that light or sound pulses at 40 Hz can reduce beta-amyloid plaque buildup in the brains of mice, suggesting resonance-based interventions could play a role in neurodegenerative treatment.

We are energetic beings first, chemical machines second. And while molecules and hormones are essential to our understanding of physiology, they are not the starting point. At the root of life is charge, field, and frequency. When this is disrupted by environmental interference—be it artificial EMFs, disordered water, or noise pollution—the symptoms may manifest as anxiety, insomnia, chronic inflammation, or poor cellular regeneration. But the cause is not always chemical. It is often electrical, quantum, and invisible to the naked eye. The tools of tomorrow's medicine will be based less on pharmaceuticals and more on resonance. Because once we understand the frequencies of health, we can tune the human body like an instrument.

9.3 Harnessing Quantum Biology for Health Optimization

If quantum biology has revealed anything, it is that the human body is not merely a bag of chemicals but a finely tuned resonant system. It responds to light, sound, vibration, and electromagnetic information in ways that transcend the linear logic of traditional medicine. With this understanding comes a profound shift in how we approach healing. Instead of focusing solely on symptom suppression, we begin to ask: how can we restore coherence? How can we help the body remember its original frequencies?

One of the most accessible tools for this is sound. Sound therapy, often dismissed as pseudoscience, is making a comeback under the name of vibroacoustic therapy and frequency medicine. Modern instruments such as binaural beats, tuning forks, and frequency generators allow practitioners to expose the body to precise vibrations. When selected with care, these frequencies can entrain brainwaves, modulate nervous system states, and even promote cellular repair. The mechanism is entrainment—the tendency of biological rhythms to synchronize with external stimuli. When exposed to healing frequencies, the body begins to resonate in harmony.

But it doesn't stop at sound. Light, too, plays a critical role. Red and near-infrared light therapy, for example, penetrates deeply into tissue and stimulates mitochondrial function through a mechanism known as photobiomodulation. The absorption of photons by cytochrome c oxidase within mitochondria increases ATP production, reduces inflammation, and accelerates wound healing. This is not theory. It is being used in elite sports, neurorehabilitation, and dermatology clinics worldwide. Yet its true potential lies in consistent, daily use for cellular optimization—not just recovery from injury.

This brings us to structured water. The idea that water can be structured—not just by nature but intentionally—offers exciting therapeutic potential. Water can be vortexed, exposed to magnetic fields, passed over specific minerals, or even exposed to healing frequencies to restore its coherence. Devices are now available that mimic these processes, producing water with altered surface tension, zeta potential, and electromagnetic signature. Anecdotal reports and early studies suggest that structured water improves hydration at the cellular level, enhances detoxification, and may even influence gene expression by improving the energetic environment of the cell.

These interventions are not random; they are based on quantum principles. Coherence, resonance, and field-based regulation form the foundation of this new paradigm. That's why advanced practitioners are beginning to use quantum biofeedback systems to measure subtle energy imbalances in the body. These devices assess electromagnetic emissions, heart rate variability, and skin conductivity to identify patterns of disharmony. More than diagnostic tools, they offer therapeutic feedback—vibrational signatures designed to restore balance in real time. They do not treat disease; they promote order.

What does it look like to live in alignment with quantum biology? It means waking with the sun, allowing full-spectrum light to reset your circadian rhythm. It means grounding your bare feet into the earth to absorb electrons and balance your bioelectric charge. It means drinking structured, mineral-rich water that feeds your cells more than calories ever could. It means listening to your body's responses to sound, light, and space—not as metaphors, but as real biological inputs. Health, in this view, is not the absence of disease. It is the presence of resonance.

And here's the deeper truth: when the body is in a state of coherence, healing is the default. Cells communicate clearly. Enzymes function efficiently. Mitochondria produce abundant energy. The immune system becomes self-regulating. It is not magic. It is physics. The laws that govern atoms also govern you. And the more we understand and respect those laws, the more we align with the intelligence built into every layer of our biology.

Quantum biology is not a supplement or a protocol—it is a way of seeing the body as a living antenna, constantly receiving, interpreting, and broadcasting information. Healing begins when we stop blocking that information and start tuning into it. We are made of light, water, and vibration. Medicine should reflect that.

Chapter 10: Epigenetics – How Your Lifestyle Rewrites Your DNA

10.1 The Epigenetic Revolution Beyond Genetics

For over a century, science has idolized DNA as the ultimate blueprint of life, the unchangeable script from which everything about us is read. We were taught that our genetic code determined our eye color, our height, our risk for cancer, diabetes, depression—everything. DNA, we were told, was destiny. But this narrative, once upheld as absolute truth, is no longer accurate. It is not false because DNA is unimportant, but because DNA alone does not explain the full story of health, disease, or human potential.

Enter the world of epigenetics. In Greek, "epi" means "above" or "over," and epigenetics quite literally refers to what sits on top of the genes. It's the layer of regulation that determines which genes are expressed, when, in which cells, and to what extent. Epigenetic mechanisms don't change the DNA sequence itself. They act more like a dimmer switch than a light switch, modulating how brightly or dimly a gene is expressed. Two people may share identical DNA, but their gene expression can diverge dramatically depending on their life choices, environment, stress levels, diet, sleep, and even the emotional tone of their relationships.

This is not philosophical speculation—it is molecular biology. One of the key players in epigenetic regulation is a process known as DNA methylation. Here, methyl groups (tiny molecules made of one carbon and three hydrogen atoms) attach to specific regions on the DNA strand, often near gene promoters.

When enough of these methyl groups accumulate in a particular spot, they effectively silence the gene, preventing it from being transcribed into RNA and thus translated into a protein. It's like placing duct tape over a line of code in a software program: the instruction still exists, but it can't be read or executed.

Histone modification is another major mechanism. DNA does not float freely inside the nucleus of a cell—it is tightly wound around spool-like proteins called histones. These histones can be chemically modified by acetyl, methyl, or phosphate groups, which alters how tightly the DNA is wound. Loosely wrapped DNA is more accessible to the cell's machinery and more likely to be expressed. Tightly wrapped DNA is hidden away, less likely to be activated. These structural changes, while invisible to the naked eye, have massive consequences for which genes are turned on or off at any given time.

Now consider this: these modifications are not random. They are responsive. They react to signals from the environment. Exposure to toxins, nutrient deficiencies, chronic stress, poor sleep, lack of movement, social isolation, trauma—all of these factors can send messages to the epigenetic machinery. The result is a shifting pattern of gene expression, one that adapts continuously, sometimes for better, often for worse.

The implications are staggering. Epigenetics reveals that our DNA is not a rigid, pre-written script, but a dynamic, adaptable library of possibilities. Our choices—what we eat, how we move, how we breathe, whether we forgive or hold on to resentment—act as editors, constantly revising the manuscript of our biology. We are not passive carriers of our genes; we are active participants in their expression.

Perhaps most shocking of all is that epigenetic changes can be passed on to future generations. This process, known as transgenerational epigenetic inheritance, flies in the face of the old genetic model which held that only changes to the DNA sequence itself could be inherited. But we now know that the chemical tags added to DNA can survive through meiosis—the process by which sperm and egg cells are formed—and can be transmitted to offspring.

This discovery has profound consequences. If a grandparent experiences famine, war, or emotional trauma, their offspring may carry the epigenetic scars of that experience, even if they never endured it themselves. Studies in mice have shown that exposure to a specific odor paired with a mild shock led to a fear response—not just in the original mouse—but in their children and grandchildren, despite those generations never encountering the odor or the shock. Human studies, such as those involving descendants of Holocaust survivors or individuals affected by the Dutch Hunger Winter, reveal similar epigenetic fingerprints, linking ancestral experience with present-day gene expression patterns and health outcomes.

The past, it turns out, is not buried. It is inscribed on our chromosomes, not as DNA sequence, but as epigenetic memory. This raises both haunting and hopeful possibilities. Haunting, because it means we inherit more than eye color and blood type. Hopeful, because epigenetic marks are reversible. We are not bound by the past if we understand how to influence the present.

This brings us to the concept of biological age, a topic that has captured the attention of longevity researchers and biohackers alike. Traditionally, age has been measured by the number of birthdays we've had. But chronological age tells us little about how well our bodies are functioning. Biological age, by contrast, refers to how old our cells and tissues actually are, and one of the most precise ways to measure this is through epigenetic clocks.

These clocks, developed by scientists such as Steve Horvath, use patterns of DNA methylation to estimate a person's biological age. The results can be unsettling. Two people of the same chronological age may differ by decades in biological terms. One may have the methylation pattern of someone ten years younger, while the other may resemble someone twenty years older. The difference lies not in their genes, but in how those genes are being expressed, shaped by lifestyle and environment.

Factors that accelerate epigenetic aging include smoking, obesity, chronic stress, environmental pollution, and sedentary behavior. Factors that slow or reverse it include exercise, plant-based nutrition, fasting, meditation, high-quality sleep, and meaningful social connections. Several studies have now demonstrated that specific lifestyle interventions—such as comprehensive diet, stress reduction, and physical activity—can actually reduce biological age as measured by methylation patterns in as little as eight weeks.

This turns the entire notion of aging on its head. Aging is not merely a matter of time—it is a dynamic process that can be influenced, modulated, and potentially even reversed. What we once accepted as inevitable degeneration is now seen, at least in part, as a reflection of cumulative epigenetic signals—many of which are within our control.

Yet this also carries a profound responsibility. If our choices can damage our epigenome, they can also protect and heal it. It means that personal health is a form of communication with our genes, a daily dialogue between behavior and biology. What we eat, how we move, how we sleep, how we love—all of it matters. We are shaping the next chapter of our lives, and perhaps those of future generations, with every decision we make.

The idea that we are victims of bad genes is rapidly losing ground. Yes, there are rare monogenic diseases where DNA mutations are deterministic. But for the vast majority of chronic illnesses—diabetes, heart disease, autoimmune conditions, depression, cancer—epigenetics plays a far greater role. And this means we are far more powerful than we were led to believe.

We are living through an epigenetic revolution, one that is changing how we think about disease, prevention, and even identity. It calls on us to become more conscious stewards of our biology. Not just for ourselves, but for our children and grandchildren. Our genes may load the gun, but our lifestyle pulls the trigger—or puts the safety back on.

In the chapters that follow, we will explore how to harness this knowledge into practical strategies. But the foundation is clear: you are not at the mercy of your DNA. You are the author of your gene expression. You are the sculptor of your biological future. You carry within you a vast, responsive, intelligent system that listens to your every breath, bite, and belief—and adjusts accordingly. The story of who you become is being written, moment by moment, in the ink of your choices.

10.2 Environmental Factors That Rewrite Your Genes

The double helix may be the architecture of your genome, but your environment is the interior designer. Your body constantly reads signals from its surroundings—what you eat, breathe, think, and feel—and then translates those cues into molecular language that reshapes genetic expression. The genes remain the same, but the epigenetic annotations scribbled into the margins of your DNA change from moment to moment. This process is not metaphorical. It's biochemical. And over time, these annotations can define whether you flourish or falter, not only in your lifetime but in those that follow.

Nowhere is this more clearly illustrated than in the field of nutritional epigenetics. The food you consume is more than calories and macros; it's code. Within every bite are compounds that can donate or withhold methyl groups, the chemical modifiers that play a central role in DNA methylation. Folate, choline, betaine, and methionine are critical methyl donors. They help shape whether a gene is turned on or off at a given moment. When these nutrients are abundant and properly metabolized, your epigenetic machinery has the raw materials it needs to make precise, beneficial edits to your genetic script.

But the modern diet, increasingly stripped of whole foods and saturated with synthetic additives, often fails to supply these essential cofactors. When methyl donors are deficient, the machinery of gene regulation becomes unstable. Genes that should be silenced may remain active, contributing to inflammation, tumor growth, or autoimmune reactions. Conversely, genes that should be expressed—like those involved in DNA repair or antioxidant defense—can be suppressed. Over time, these epigenetic misfires accumulate, not because your DNA is faulty, but because the information your cells receive from food is garbled or insufficient.

And then there's stress—both psychological and physiological. While we often think of stress as a mental state, the body interprets it as a signal that shapes its deepest biology. When exposed to acute stressors, the body mounts a predictable cascade of responses: cortisol rises, heart rate accelerates, inflammation increases. This short-term adaptation is useful when facing immediate danger. But chronic stress, the kind born from financial insecurity, toxic relationships, social isolation, or childhood adversity, imprints lasting marks on the epigenome.

These marks are not just theoretical constructs—they're visible, trackable changes in methylation patterns and histone modifications that alter gene expression for decades. For instance, studies have shown that individuals who experienced significant trauma early in life display altered methylation of the glucocorticoid receptor gene, a key regulator of the stress response. This epigenetic scar can persist into adulthood, making them more reactive to future stress, less resilient, and more susceptible to anxiety, depression, and even metabolic disorders. Their cells, in a way, carry a memory of past pain that reshapes their biology in the present.

Perhaps even more striking is that these changes don't stop at the individual. They can pass on to offspring. Animal studies have revealed that maternal separation, a proxy for early-life trauma, produces epigenetic modifications in brain genes involved in stress regulation—not just in the mother, but in her pups, and even in subsequent generations. In humans, the children of individuals who lived through genocide, war, or severe famine often show similar alterations in stress-related genes. The wounds of history are etched into the genome, not through mutations, but through epigenetic signals carried silently across time.

But stress isn't the only environmental insult that rewrites our biology. Toxic exposures do so as well—silently, persistently, and often irreversibly. Heavy metals like mercury, cadmium, arsenic, and lead interfere with normal methylation processes. They can promote aberrant DNA methylation, silencing tumor suppressor genes or activating oncogenes. Pesticides, which saturate much of the modern food supply, have been shown to produce transgenerational epigenetic changes, increasing the risk of obesity, reproductive disorders, and neurodevelopmental delays. Endocrine disruptors like BPA, phthalates, and parabens alter hormonal pathways not by mimicking hormones alone, but by modulating the expression of genes involved in hormone synthesis, receptor sensitivity, and detoxification.

What's most insidious is that many of these compounds don't need to be present in high concentrations to exert effects. Because they interface with the epigenetic system, their impact can be disproportionate to their dose. Small, repeated exposures—through plastic containers, air pollution, cleaning products, or cosmetics—create a cumulative epigenetic burden. It's not simply about what enters your body in large quantities, but what whispers to your genes often enough to alter their tune.

This means that health is not just a matter of avoiding genetic "bad luck," but of managing a daily dialogue with your environment. Everything from your morning coffee to the type of light you use at night, the air you breathe, and the emotions you feel—these are all epigenetic signals. And they are either shaping your biology toward resilience or toward dysfunction.

10.3 Optimizing Your Epigenetic Expression

If your environment can harm your epigenome, then it also holds the power to heal it. This is the promise of modern epigenetics—not only as a diagnostic tool but as a new frontier in proactive medicine. While genetics offers limited intervention opportunities (you can't change your inherited DNA), epigenetics is inherently malleable. It's a system designed to respond to inputs. The key, then, is learning how to provide the right ones.

Lifestyle medicine is the foundation of this approach. It is not a collection of "tips" but a disciplined strategy for signaling safety, nourishment, and coherence to your biology. At the heart of this is diet. Whole foods—especially those rich in methyl donors and bioactive compounds—are not only nourishing; they are regulatory. Leafy greens like spinach and arugula, cruciferous vegetables, berries, and sulfur-rich foods like onions and garlic provide the very substrates your cells need to perform clean, efficient methylation. Polyphenols in turmeric, green tea, olive oil, and pomegranate have been shown to demethylate silenced tumor suppressor genes and modulate inflammation-related pathways.

Exercise, too, acts as a genetic editor. Physical movement doesn't just burn calories—it sends epigenetic messages. Studies have shown that a single bout of aerobic exercise alters the methylation status of genes involved in energy metabolism, inflammation, and insulin sensitivity. Regular exercise essentially reprograms the body's response to aging and disease. The epigenome interprets physical activity as a sign of vitality and adapts gene expression accordingly, reinforcing strength, endurance, and cellular repair.

But all of this is undermined if stress remains unaddressed. Chronic stress dysregulates the HPA axis and floods the body with cortisol, altering gene expression in the hippocampus, prefrontal cortex, and immune system. Meditation, breathwork, and consistent sleep not only lower perceived stress—they actually reverse stress-induced methylation changes. Mindfulness-based stress reduction techniques have been shown to modulate expression of genes involved in inflammation and cellular resilience. In essence, a calm nervous system writes a different genetic story than one plagued by anxiety and overstimulation.

Supplementation offers another layer of support—especially for individuals with genetic variants that impair methylation, such as MTHFR mutations. In such cases, targeted use of methylated B-vitamins (like methylfolate and methylcobalamin), along with cofactors like magnesium, zinc, and trimethylglycine, can help restore proper methylation dynamics. These nutrients are not to be used casually; too much methylation can be as harmful as too little. The key is balance, informed by testing and clinical insight.

Speaking of testing, we now have the tools to peek into the epigenome with unprecedented clarity. Advanced tests like DNA methylation panels and biological age clocks allow us to assess where our genetic regulation is faltering—and where it's thriving. These tests measure which genes are being silenced or expressed, and whether your biological age aligns with, exceeds, or lags behind your chronological one. This information allows for precision health interventions, tailored not just to your symptoms, but to your molecular terrain.

With this data, a personalized epigenetic strategy can be constructed. If your methylation profile suggests accelerated aging, interventions may include time-restricted eating, sulforaphane-rich foods, NAD+ precursors, and strategic fasting. If inflammatory genes are overexpressed, protocols might prioritize antioxidant therapy, detoxification support, and omega-3 supplementation. If sleep genes are dysregulated, light exposure, magnesium, and melatonin precursors may be emphasized.

The future of medicine lies in this intersection—where genomics meets lifestyle, where data meets daily practice. It's not about silencing symptoms but about retuning the entire biological orchestra. The epigenome is listening, always. Every meal, every step, every breath, and every thought is a note in the symphony of your health. The question is: what kind of music do you want to create?

This vision of health isn't passive. It requires attention, intention, and consistency. But it is also profoundly empowering. You are no longer a slave to your genetic inheritance. You are the conductor of its expression. What you choose to do today doesn't just affect you—it shapes the health of your future self, and possibly the generations that follow. In this sense, every choice becomes sacred. Every moment is an opportunity to write a better biological future—starting now.

Conclusion: Reclaiming Your Health Sovereignty

The greatest wealth is health, but perhaps the deeper tragedy of our time is that we've forgotten how to be healthy. Not because the knowledge doesn't exist, but because it's been obscured—buried under decades of pharmaceutical reductionism, siloed specialties, and a system more focused on managing disease than cultivating vitality. What you've read in these pages is not merely information. It is a reclamation. A restoration of truths that were once intuitive, lived, and embodied—truths now resurfacing through science, tradition, and your own direct experience.

Throughout this journey, you've been reintroduced to the intricate symphony that is your body. A being made not only of organs and cells but of fields, rhythms, pulses, and feedback loops. A system so complex that no single pill, protocol, or practitioner can fix it from the outside. Yet it is also so intelligent that, when supported correctly, it can orchestrate healing with precision far beyond what we've been led to believe is possible. The most radical act, perhaps, is not in hacking your biology, but in learning to listen to it again.

Health sovereignty begins with knowledge, but it matures through action. You now possess the frameworks that were never taught in school, rarely discussed in medical offices, and too often dismissed by modern institutions. These frameworks begin not with treating a symptom but with identifying which system is out of tune. Is it your microbiome? Your circadian clock? The flow of your lymphatic fluid or the density of your mitochondria? Is it the texture of your fascia, the rhythm of your breath, the light entering your eyes, or the unspoken grief altering your genes?

To understand health in this way is to view your body as an ecosystem. You are not a collection of isolated parts, but a unified whole that functions in concert. Every function is interdependent. Every dysfunction is systemic. And every healing process is a remembering—a return to balance, rhythm, flow, and coherence.

What the chapters of this book have done is peel back the layers of that ecosystem, revealing what's been hiding in plain sight. Your gut is not just a digestive tube—it is a sensory, hormonal, and neurological command center. Your mitochondria are not simply energy factories—they are environmental sensors that respond to light, toxins, and electromagnetic fields. Your fascia is not inert tissue—it is your internal internet, transmitting messages faster than nerves. Your lymphatic system is not just drainage—it is immune intelligence in motion. And your DNA? It is not a rigid blueprint, but a responsive manuscript, rewritten daily based on your thoughts, foods, exposures, and relationships.

This is not idealism. It is biology—just not the kind that gets airtime on TV commercials or fits neatly into fifteen-minute doctor visits. It is the kind of biology that empowers you to take radical responsibility for how you live. The kind that reminds you your symptoms are not enemies, but messages. That fatigue is not laziness—it's a spark being smothered. That inflammation is not just a chemical reaction—it's your body crying out for realignment. That depression is not just serotonin deficiency—it's disconnection—from nature, from rhythm, from meaning.

And now you know how to reconnect.

You know the light you expose your eyes to first thing in the morning is not trivial—it programs your hormones, your metabolism, your cognition. You understand why sleep is not a passive activity, but a time when your brain detoxifies itself through the glymphatic system, when your fascia regenerates, your cells repair, and your epigenome resets. You recognize that the food on your plate is not just nutrition—it is molecular instruction. That processed sugars, seed oils, and synthetic additives are not neutral—they are agents of dysfunction, misinforming your biology every time you ingest them.

You've also seen how movement is more than fitness—it's information. That rebounding can stimulate lymphatic flow. That spirals and waves in motion reawaken fascial elasticity. That walking barefoot isn't "woo"—it's bioelectric recharging. That cold plunges and heat exposure are not fads—they are invitations to remember what your mitochondria evolved to expect.

Perhaps most crucially, you've encountered the ways in which the invisible governs the visible. You've seen how electromagnetic fields, sound frequencies, and coherent water shape the very structure of your cells. How quantum effects operate not only in plants and birds, but in your enzymes, your neurons, your DNA. And how these discoveries are not futuristic—they are foundational. They remind us that our biology is not mechanical, but energetic. That our healing is not chemical, but electrical. That the field—not just the molecule—is where transformation begins.

And yet, knowledge alone is not enough.

Health sovereignty is not a passive awareness—it is a lived practice. It is choosing to stand in your own authority, to listen to your own signals, to move at the rhythm of your own biology rather than the artificial tempo imposed by society. It is understanding that while experts can guide, only you can embody. That while protocols can support, only you can sustain. That while supplements can assist, only daily choices can transform.

This doesn't mean perfection. It means participation. You will not always eat clean, sleep eight hours, or wake with gratitude.

But now you know why it matters. You know what your biology needs, and what it doesn't. You know that health is not about avoiding death—it's about amplifying life. About clarity of thought, strength of body, steadiness of mood, sharpness of intuition. And you know that these are not gifts for the lucky— they are the birthright of the informed.

In claiming this birthright, you become dangerous—in the best possible way. You become harder to manipulate. You become less dependent on pills, procedures, and protocols. You begin to question what is sold as normal, from blue light at midnight to processed meals in hospitals. You begin to feel what vitality is supposed to feel like—not stimulation, but coherence. Not numbness, but presence.

This is the revolution. Not in the streets, but in your cells. Not through protest, but through practice. The revolution in health is quiet, persistent, sovereign. It is you turning off the Wi-Fi at night. You stepping into the sun without sunglasses. You stretching your fascia before opening your laptop. You drinking structured water. You choosing joy over obligation. It is not complicated. It is just unfamiliar—because you've been taught to forget.

But the forgetting ends here.

You've walked through the corridors of the human system— through the gut and the glands, the fascia and the fields, the genes and the geometry. And now you stand at the threshold of application. This is not the end of a book. It is the beginning of a practice. Of thousands of small decisions that add up to a reclaimed life. A life where the light you wake to matters. Where your breath has rhythm. Where your food is alive. Where your movement is intentional. Where your sleep is sacred. And where your health is no longer outsourced.

The missing truths are missing no more. They are yours now—living within you, waiting not to be memorized, but to be lived. The revolution in health does not belong to doctors or gurus. It begins with you. With the sovereignty to say: I am no longer waiting for someone else to fix me. I am re-entering partnership with my body. I am listening to the subtle signals. I am restoring the rhythms that make life whole again.

Your biology has been waiting for this reunion. And now, it begins.